Empi

Empire of Scientism

The Dispiriting Conspiracy and Inevitable Tyranny of Scientocracy

James Tunney

James Tunney is the author of this book and owns all copyright © in this work and assert all moral rights thereto 2021.

Cover image and inside art copyright © James Tunney.

ISBN: 9798721894770

To Katie and Mary

Easter Monday 2021

Also by James Tunney

NON-FICTION

The Mystery of the Trapped Light – Mystical Thoughts in the Dark Age of Scientism

The Mystical Accord – Sutras to Suit Our Times, Lines for Spiritual Evolution

FICTION

Blue Lies September

Ireland I Don't Recognise Who She Is

Contents

Introduction: Central Cause for Concern 9

The Spectre of Scientism.................................. 15

The Emergent Empire of Scientism 19

The Threat of Digital Tyranny 23

The Foundation of Fear 27

From Machine Mind to World Government 33

Persecuted Rebel Myth of Faustian Scientists... 40

Loss of Liberty and End of Freedom................ 45

Controlling Human Consciousness 49

Submission and Acquiescence to Scientism..... 57

Scientocracy or Technocracy Beyond Regulation............ 62

Apprehension of Technocratic Tyranny 66

Away From Dispiriting Conspiracy 76

Towards the Animasphere................................ 84

And What If Then?.. 100

About the Author... 101

Anatomy of Insight, oil painting by James Tunney.

Introduction: Central Cause for Concern

This short work has grown out from a growing sense of concern at the direction we are heading. It derives from my knowledge and work in law and regulation and a family interest in politics. It is informed by my work as an artist and writer and studies in mysticism. As the positive, light side of my poetry and mystical exploration evolved it created a sharper contrast with my apprehension about the nature of our governance. My perception of the need for spiritual evolution and inner development has both revealed and made starker some deadly threats to our consciousness and very existence as humans. Mystical perception may also make certain forces more manifest.

I remember long ago in a London pub, a stockbroker came in and put his mobile phone on the bar. That was the first time I saw one. He was the only person with such a big contraption and was thus an object of attention being exceptional. Now I am regarded as eccentric because I have so far managed to avoid having one. Because of mathematical, economic, commercial and technological forces creating networks, anything which can be linked thereto will be. People are also regarded as things by many. Novelty, perception of opportunity, choice and propaganda associated with perceived benefits of networks make people blind to the equally big costs. The huge hi-tech front of fun, entertainment, education and convenience has an equally gigantic back of addiction, entrancement, enmeshment and control. Advocates of computers advertised their policy of getting kids 'hooked' thereto early on. We are all hooked, netted and caught in a web of networked machines now. The facade of fun is giving way very rapidly to the fact of surveillance, angst, anxiety, anti-socialness and worse. What could be worse? Well… we must realise that a tyranny has formed through the illusion of convenience, market mesmerism, opportunities for control and our own fear and foolishness. The tyranny is co-ordinated because all such networks need to be so at the very least to function as infrastructure.

Independently of commercial forces, whether from capitalism or communism, there is deeper concentration or co-ordination. The base philosophy of materialism, as sharpened by commerce, computerisation and technology, causes and reinforces an ideological interest in global governance through science and technology. Cybernetics is about governance, as is much network development. Such technological governance is being assumed by those who benefit, advocate and implement it. That co-ordination occurs is plain to see. An ideology that assimilates elements from a number of forces, but mostly 'scientism' is leading to an ideocracy, more specifically a 'scientocracy.' This means that governance by this political ideology, dogma and all-encompassing mind-set can occur. Governance must be global now it seems. The ideology that causes co-ordination dismisses alternative conceptions to the rise of global tech-governance or what I term 'globetechgov.' Previous ideologies based on being Marxist or capitalist are disregarded in favour of collaborative control based on submission to, and dependency on, technological networks. The interests of the governing control-apparatus in what is an emergent empire is already triumphing over competing interests such as freedom of expression.

This is not an argument against legitimate public health measures or an engagement in etiology of recent events. Nevertheless, fearful conditions produced by present medical and public health circumstances have led to minimalist protection of individual rights. Sovereignty has been ceded to centralised, bureaucratic control supposedly calculated to enhance the common good in a collectivist endeavour run by specialists. Such experts have assumed an authority which is impossible to question, having the status of unassailable truth. They are free to operate according to an agenda advocated by them without any real, democratic restraint. The agenda is articulated in terms of one chosen individual interest as defined solely by perceived public health risk presently. Further solutions emerging through the propaganda model, all assume a greater submission to international surveillance webs. Choice is eradicated in favour of dictate. Further erosion of democratic control occurs, shifting to unelected overseers who can create coercion. So people have been isolated. Children have been forced to watch screens for interminable lengths. Movement has been severely restricted and so on. Whatever justifiable public health consideration may have been utilised, actions have been grossly disproportionate to risk. The technological networks have become more autocratic and entrenched. Competition has been reduced. Economies have been decimated. More conditions for global centralisation of power as a supposed solution have been established. Everything is made into nails for hammers to hit.

I recall my sense of shock when someone for the first time took out a mobile phone at a dinner party and communicated with someone else not there. I thought it exceptionally rude. I remember when first someone fact-checked a statement I made with a mobile in a social context and was amazed at such audacity and assumed authority of information found whilst being made aware of their lack of confidence in my observation and obvious incapability to assess it on their own. I still find it difficult to see younger people resorting to a mobile as a device to cover for their lack of social skills that have been partly undermined by their existing commitment to 'social' media. Now, prospects of mobiles being replaced by wearables or implants have increased dramatically. The lone implantee is like the first mobile phone user I saw. I hope as well to be one of the last without an implant. Unfortunately choice disappears. You are now entering the Empire of Scientism and have also become a subject or more realistically a slave. Unfortunately again, there may be no place to escape to. That inability to escape individually is becoming the fate for the majority of our species.

The centralising or centripetal force of governance at global level through technological networks certainly comes from psychological tendencies to control. From the dawn of civilisation, human groups sought to control other humans. The science of cybernetics, developed by Norbert Wiener and others, is about study and application of control systems and communication in animals and humans. This involves a complex comprehension of information flows, feedback and communication systems. Cybernetics, assisted by AI, surveillance and network dependency, create the tools for global governance. Cybernetics is governance. The means is the end. Realising potential actual control of the human mind drives those who will take governance reins without constraint. The possibility is facilitated by continual severance of democratic nexus between the controlled and controllers. Administrative and legal sovereignty shift to international bodies without any genuine oversight or constraint. Accumulation of power in institutions committed to steering governance means that national sovereignty is diminished. Now, institutions who have acquired such power accuse national politicians of not having the desire to govern, after having magnetised power to their domain.

Networks create other networks. The emerging global governance system will be like a machine. The machine will constitute a system which reinforces its own nature in a recursive way. It will act reflexively on humanity transforming us to correspond to itself. Machines and computers operate reflexively on people. The machinery will govern governments and treat problems as mechanisms, systems and objects. Humans will be treated as objects subject only to objectives and not be regarded as subjective forces with selfhood and human dignity or be unique or special. An elite caste beyond control will operate the system for its own benefit. That is the history of power structures. The Empire of Scientism will constitute governance of the machine, by the machine, for the machine.

As an antidote I argue for an 'animasphere.' Such concepts may operate as a system to repair in interstices between broken fragments of our human heritage. The fractured bowl of our common experience has left shards and splinters that can sever and wound in their sharpness. The Japanese 'kintsugi' repair mode refers to an approach that seeks to emphasise rather than deny cracks. Golden repair may underline broken places by having obvious concentrations of repairing substance, usually gold or silver. Repair becomes re-constitution and enhancement. We need a kintsugi approach to conceptual and spiritual cohesion within the human race. Otherwise we will be hoisted on intellectual gambrels of authoritarianism. John Maynard Keynes pointed out that authoritarians act out ramblings of some dead academic scribblers. Scientists must be more careful when making sweeping statements in a surging sense of superiority deriving from how good science informs and performs. Their extrapolations may mutate virally in other brains to justify horrors unintended.

Thus we have grave cause of concern, not confined to contemporary circumstances. Scientism arises primarily from combining ways of thinking and being with dismissal of others ways. Scientism derives from a tendency within science and is a product of power realised by magical success from some methods therein. This thinking creates beliefs about order, systems and standards and can be allied with instruments to entrench and extend its apparatus. The machine will be run by a class convinced by such instrumental power and conditioned by data omniscience. The linked security-industrial-military-pharmaceutical complex is complemented and will be superceded by a network of politically-remote control. Humans become things in this 'internet of things' and liberty disappears as will the human race as we know it. Tech-Totalitarianism with biosecurity is here.

The Spectre of Scientism

"More and more the world may be run by the scientific expert."

Bernal, J.D.
The World, The Flesh, The Devil:
An Inquiry into the Future of the Three
Enemies of the Rational Soul (1929).

Mary Shelley sought to warn the world and so do I. Frankenstein's monster represented science itself but also indicates a dangerous sort of credulous idol worship. We live in times defined by the ideology of 'scientism' and face a future where it will be elevated into an extensive empire of machines that controls the world. We are shaped increasingly by the nature of machine structure at the expense of nature itself and our nature. *Scientism involves the application and misuse of scientific knowledge beyond an adequate and appropriate limit, in an exclusive way, such that it may become a dogma or ideology.* Scientism elevates science beyond its domain of validity, creating a dominating and destructive force convinced of its own correctness without qualification. Scientism denies that which cannot be proven even when science has not sought nor examined evidence due to absolute certainty in its method, despite constantly altering and adapting itself. Scientism culminates in an idolatry of matter and force supported by undue focus on methods and techniques from theories in empiricism, positivism, physicalism, naturalism, materialism and mechanistic thinking without respect for their limitations. Having crystallised in an

Enlightenment-rejection of divine authority, scientism now promotes exclusive authority with a zeal matched by all other fundamentalisms despite obvious contrast with its own claims about neutrality and objectivity. Scientism (and even science itself) denies many problems, predicaments, catastrophes and cruelties created in its name while advocates constantly criticise other systems perceived to be in competition for dominance. Scientism is ideology or religion while machinery and technology are idols to worship. Logos is the difference between ideas and ideology and that is now supplied by the idols and sigils of commercial logos as pale substitutes.

Belief in science's superiority, beyond its method, creates a philosophical conservatism paradoxically characterised as progressiveness. Scientism is spawned by control of instruments that serve powerful institutions creating beneficial conditions for personal adaptation to its apparatus by subscribing thereto. Scientism is related to the idea that society and humans are machines or, insofar as they are not yet, will become so. From Bacon, Newton, Descartes and others comes the idea of a machine or clockwork universe. From Julien Offray de La Mettrie with *Man a Machine* (1747) to Marvin Minsky comes the belief that we are machines. Scientism wants to make the model replace complex reality as thinkers like Owen Barfield identified. The undeniable power of scientific systems and technology manifesting in corporate might, controlling markets and capturing institutions, means that scientism will govern the globe. We are seen as machines to be run by machines. A strange spell of fascination with shiny new technology combines neophilia, technophilia or even mechanophilia as accelerators for a technosphere forming a new empire. As previous empires changed from control of resources to finance to digital economy to information, the ultimate empire seeks to manage everyone mentally. This empire of empires, or empire of the mind, will be organised as a machine by social and actual machines in order to manage and transform us into machines within those machines. Scientism is intellectual glue that makes this empire cohere. Technology, digital networks and cybernetics are scientific offspring that allow claims of power seem almost divine, true and valid thus creating a sense of inevitability and reinforcing certainty. Unfortunately, we are the ultimate colony to be plundered. A zombie version of us may be tolerated by our masters within a technosphere they create, a drugged-up simulacrum in a disorientated soporific domain. But the import of current trends is even worse. We need to realise our predicament, wake up, unhook, critically examine, re-discover and utilise our power of spiritual consciousness. Scientism ideology combined with technological networks and the nature of administration by bureaucrats, create conditions for tyranny. *Scientocracy may be defined as the primary organisation of society by scientists, technical support, managers of technological networks and others using or purporting to use science as the method of total, global governance.* While some advocate it, I argue that it must be totalitarian, as C.S. Lewis and others do. Postman refers to 'technopoly' in *Technopoly: The Surrender of Culture to Technology* (1992). For Postman, technopoly refers to advanced management of society more pronounced and totalitarian than mere technocracy. Technocracy is used to define the serious threat by Wood in *Technocracy Rising: The*

Trojan Horse of Global Transformation (2014). All such critics indicate that scientific method will be projected as means of arbitrary global management.

Scientism is one cause of authoritarianism. From Bacon and the Enlightenment came the idea that if one measures, quantifies, predicts and is objective, empirical and efficient, we can obtain truth based on observation, enquiry and laws. Such truth is genuine in a way superior to the imaginal, supernatural, theological, cultural or any non-verifiable source, supposedly. Displacing other ways, limitations are forgotten. By concentrating on cause and effect, predictability and mathematics, one seeks truth even in domains beyond natural sciences. Knowledge and information are king and acquire a life of their own whereby uses can be applied for purposes beyond the strict scope and methodological reliability in original inquiry. That about which we have no information, data nor intelligence is regarded as useless, valueless and non-existent. Machine-mechanism requires functionality, demonstrates use and manifests tangible results. That which cannot be tested yields no results and being beyond test methods is not really real. The invisible, immaterial, supernatural or natural incapable of reduction or detection, are thus disregarded. Liberation from not investigating that which cannot be tested and engagement in meaning or metaphysics makes more time to spend on the tangible and visible. Undue focus on the controllable creates more control and spins a vicious circle through obsessive quantification. A dangerous illusion of complete insight creates undue certainty if applied inappropriately. Tyranny of experts in easy association with authoritarianism often reduces rights rather than improve them.

The Emergent Empire of Scientism

Human history tells of constant conquest and colonisation. It is crystal clear over the last generation that the force of technology and authority of science has come to dominate our world. We are becoming addicted, dependent and hooked to a system that controls us and accumulates power over us every day. When the world's electronic nervous system as explained by Marshall McLuhan in *Understanding Media* (1964) assimilates us, autonomy is relinquished. The end will be an entire empire where we are slaves or servants at best. What I term 'The Empire of Scientism' is finally erecting a system of tyranny as proposed nearly a century ago, as dreamt by despots, fortunately for us hitherto without technical means to achieve it. *The Empire of Scientism refers to an emergent system of global governance based exclusively or excessively on that dogma that will necessarily be totalitarian.* Whether it is intentional, calculated, coordinated or not is almost irrelevant to the emergence in fact. The pattern in apparent chaos in centripetally aligning governance is clear, except to people who choose not to see, oblivious to the obvious. Enlightenment materialism, industrialism and imperialism have encouraged a movement of scientific naturalism that sought to colonise all knowledge and ultimately the human mind and spirit. We are encountering a present, and facing a future, involving permanent cultural revolution in a transition and likely end of humanity. The human is moving irresistibly towards the machine and being assimilated by it. The idea that humans actually are machines is now aided by argument that we must become more machine-like and less biological. Elon Musk points out that because of mobile phones we are cyborgs. While exploring transhumanism he, like Stephen Hawking, warns that AI will control and perhaps destroy us. We are going to transition from mere machine dependency to become machines. Transition will not be a matter of choice insofar as established networks want everybody linked. Networks and webs entrap us with propaganda of convenience. A new honeypot of habits has made us ready for servitude. Commercial, corporate and communist control can combine cyber-networks to make management of people much easier. The most probable outcomes look like a prison, zoo, slavery or extinction of humanity much quicker than by any natural catastrophe. As we cede sovereignty daily, we lose opportunity for escape from a vortex of technological dependence and control. We are wandering into scientocracy, technocracy, technopoly or a technetronic society, as control increases exponentially. Vehicles of social media on superhighways attract a tech-rule often regarded as an ideology also. Managerialism in Silicon Valley or the Silk Road merges with technological administration. Burnham in *The Managerial Revolution* (1941) and Enteman *Managerialism: The Emergence of a New Ideology* (1993) indicate how a complementary ideology reinforces the likelihood that technocrats assist formation of a totalising governance system. In addition there is an associated cult of the expert. Experts work with authoritarian administrations in the context of foreign aid. This has been identified by William Easterly in *The Tyranny of Experts: Economists, Dictators and the*

Forgotten Rights of the Poor (2014). He identifies how policy is often based on purely technical solutions while ignoring the politics of autocracies. Thus there is a 'technocratic illusion' of a benevolent dictator who will implement technical solutions to operate in the absence of an apparatus of rights.

The century-long dream of technocratic rule over docile humans or sub-humans relieved of pretensions about deeper domains of dignity, consciousness, self and personhood, is coming true. Far superior for our masters than messy military engagement is mind manipulation and management by experts and machines. The Empire of Scientism has an inchoate scientific-religious dogma with technological idols to worship. It has a dedicated corporate priesthood whose personal benefit and desire for power incite pursuit of even more. Control by puppet-masters, with obsessive curiosity for tool potential, economic reward and institutional adulation, promote this force. While some may welcome the tech-admin, new world order of predictability in a straitjacket simulacrum of safety, it cannot be but a dreadful digital despotism. We must realise we are hypnotised, beguiled or duped into an illusion, maya or matrix by our desires. The fashionable idea that we may presently live in a computer simulation is an accurate reflection of a generation anticipating a reality they are primed for. Just as we are told religion is the biggest cause of war despite historical evidence to the contrary, so might we be primed for division fomented by the very forces proclaiming such propositions. No group, nor single person, are to blame for this but rather the ideology of scientism. Scientism has encrusted on the surface of diverse forces characterised as being a benevolent domain of science, rationality and reason fenced off from its own horrors and the 'imp of the perverse' within, that Edgar Allan Poe alerted us to. An ideological machine with hi-tech instruments and an undemocratic governance mechanism will manage us without our consent. The Empire of Scientism will constitute a massive machine to manipulate and coerce many by few in an inexorable tyranny. Machines plus mind-set will make this Empire manifest in a scientocracy, technocracy, technopoly or technetronic society that will be totalitarian.

There are philosophical critics such as Paul Feyerabend who criticised the idea that science should have a special position in society. He argued that many advances in science did not come through a special method. He claimed that there was nothing inherent in science that was necessarily liberating now as opposed to when it challenged the dominant religious ideology. Science today alienates heretics, has unquestionable dogmas and a mythology. Feyerabend also questioned the absolute validity of falsifiability. So even without limiting my critique to scientism, Feyerabend sees the ideology or fairytale that is science as capable of being an imprisoning ideology. But nevertheless, it is clear that institutions of science, especially with the ideology of scientism, worship and control of technology, facilitated by a culture of technocrats, technocratic solutions and over-reliance on experts, create conditions for an Empire of Scientism calculated to attain global governance. Absent any ethical base we will be tyrannised by objectives which suit managers and manipulators in the post-democratic system defined by planning and predictability.

The Threat of Digital Tyranny

Digitisation created the possibility of convergence. Systems that were separate came together. Power shifted to networks. People involved in such convergence became powerful. There is often an advance in science or technology which then makes scientists cautious afterwards. Curiosity never seems to hold them back. Ostensibly benign benefits of information technology blinded us to costs despite evidence of how Nazis used basic systems. The means to control have created a belief of entitlement to govern. Without democratic mandate, billionaires propose technical solutions based on belief in an absolute truth inherent in their insights. Technological convergence creates commercial concentrations of power and an ability to exploit citizens. In the absence of competition control globally, corporations are more powerful than nations. Commercial power controls communications networks by the propaganda model and exercises restraints on citizen rights. Government is then effectively exercised at places like Davos, Bilderberg, Bohemian Grove and in basements, golf clubs and other places we do not know, as well as in other Non-Governmental Organisations and influential commissions. Techniques of scientism are ideal for absolutism.

Tyranny through scientism is becoming manifest around the globe. As Jerry Muller's book *The Tyranny of Metrics* (2018) makes clear, costs of measuring outweigh benefits sometimes, resources go from frontline to technocrats, statistics are played, action is distorted, judgment based on experience declines and many important qualities are not quantifiable. Concerns seem corroborated by a descent into an apparently inescapable web and net of scientism consistent with a move towards scientocracy. Our popular channels of communication are complete now with new formulae, designations, graphs, signs and terms that were alien a year ago, always assumed to be authoritative and based on sound assumptions. Censorship from corporations that exert control over national discourse with political perspective creates deep suspicion. Pseudo-science seems triumphant in the dominant narrative. Assumptions underlying mathematical models of public policy are often beyond question despite the necessity of robust assumptions for statistics to be meaningful. Debate is immediately met with claims of authority and veracity beyond question more peremptory than religious authorities once predominant in Europe and still prevailing elsewhere. Left and right seem united in their drive towards an exclusively materialist, physicalist view of reality with other concerns ignored. Corporate pharmaceutical, medicalised solutions are sold as panaceas whilst disregarding enormous problems of such approaches. Pharmaceutical and tech-network power combine with political power utilising purified pervasive propaganda to perpetuate a new order of repression with devastating potential. Previous conceptions of privacy and personal choice are jettisoned to make mass systems more convenient for those who exercise power. The machine will get what the machine wants. Needs of the machine are greater than human needs. Humanity is told by technocrats, entrepreneurs and advocates of the emergent control system that they must just accept this inevitable relinquishment of power to the machine. The inconvenient idea of individual dignity that derives from a long Judeo-Christian and Greco-Roman trajectory, inspiring humanism as well as great traditions inherent in say Buddhism and Indigenous thinking, are rejected in favour of a notional collectivist concern. Both materialist capitalism and communism can combine in this collectivist control system made available by centralised hi-tech. Promoters at the top of the pyramid wield power we yield. We can sense something is badly amiss. But instead of seeing sense there is a major tendency to 'make sense.' Furthermore, obvious sense is not being seen. The protective apparatus around a coherent notion of freedom protected by law is being dismantled and is unlikely to be easily re-assembled absent key parts. Weapons and tools of surveillance, in our own hands and beyond underline our lack of real freedom lest we be under any illusion or have hope in benign protectors in the system somewhere. Tools we are told are exclusively for our benefit turn out to be calculated to constrain us in addition to other costs. Democracy with some local connection to political power has been undermined by constant power accumulation in institutions beyond local and national control. We are now on the threshold of a rapid movement to transhumanism and posthumanism. Technological proliferation represents an existential threat to humanity itself. Mainstream theorists explain how exponential growth means that we must just yield the human race and pass under the technological yoke. Meanwhile the 'propaganda model,' as explained by Herman & Chomsky in *Manufacturing Consent* (1988), permeating public space, distracts with multiple strategies of meaningless engagement and phoney debates, spurious controversies and events. Imperial conjurors wield illusion to attract us as our pocket is picked and our straitjacket fastened.

Before most people can comprehend transhumanism and posthumanism, save for slick marketing and sensational advances promised to cure the sick, we will constantly be turned in a washing-machine of social revolution. This trance-humanism stage is characterised by a state of deep confusion in order to condition the hypnotised public for instruction on the new order to be initiated. That new world order specifically proposed by science lovers is one of de-humanising technological control without any special treatment for humans. You are just a disposable, flawed meat-machine. Possessing knowledge, means and power of compulsion over human behaviour in evolving technocratic systems, the Empire also has an apparently enlightened and cohesive drive. With intelligence about human behaviour available, the emerging Empire can manipulate public fears whilst appearing friendly. Worse it can convince participants about rightness and nobility of power's pursuit just as all despots delude. Scientism is black magic for a new priesthood. Wands remotely control us. We serve a network we were convinced was serving us. Our attention was the product and now it will be our very minds and souls. Devices become vices via a sticky web and trapping net with chain links like Trojan horses hacking our system. Opened up and played, we are susceptible to management by fear. With fear of God lost, fear of humankind replaces it. Fear becomes the emotional foundation on which the invisible Empire of Scientism is erected.

The Foundation of Fear

The sickest thing about this slick operation will be that it is conducted on us with the appearance of noblest motivations. Liberty's end and tyranny's entrenchment will be promoted by the greatest apparent concern for our welfare. We will be told about the greatest dangers. We will be instructed in the most convincing way about inevitability of catastrophe. We will be made weary with strange coincidences of fantastic forces that have uniquely combined to necessitate relinquishment of our selfish, local and superstitious desires for sovereignty. We will be presented with our saviours in science whose sole desire is to rescue us from the horrible fate that they have selflessly identified as ineluctable. It may be one unavoidable, terrible thing after another, from asteroid to angry aliens. They will all have this characteristic in common. They will all represent an existential challenge to the human species and even the planet itself. They will all require unique and selfless talents and experience of those people who have warned us of such sudden, desperate and imminent dangers though they fail to warn us about other real threats. People who rejected strange ideas will remarkably embrace them and scorn those who become suspicious. But even without impending doom more characteristic of science than religion now, the future even without catastrophe may be a worse disaster for humanity anyway. We are told that we must give in, give up, give way, give away and surrender even our minds and body to technology and ringmasters that orchestrate the charade of public affairs with diminishing pretension of genuine interest in our human condition. The enterprise of science, if allowed to continue in its present trajectory, will make most of us slaves. Don't take my word for it, just study what some of the greatest and most influential, celebrated scientific advocates and more particularly scientism promoters, have stated. Science and scientism are about power and control over the material world and ultimately the mind of humankind. Despite undoubted benefits of progress there lurks a pathological presumption, somewhere between problematic and psychopathic, in relation to its unique role. Amoral arrogance allows those associated with inflation in scientific knowledge and confusion of certainty with truth in relation to more complex conclusions, assume for themselves and their clusters a self-assigned ability to control those lesser mortals burdened down with superstitious consciousness and associated restraint. We would do well to fear disciplines whose highest lights proclaimed the need to torture or torment nature to squeeze secrets out and indeed practice this while constantly re-iterating historic wrongs of other institutions it opposed for doing similar things on a far smaller scale.

You may hope that clarity will come from people who are paid to think. People who study philosophy and humanities will surely seek a safe human harbour where enlightened reasoning and debate occur in a sheltered place? Alas, dominance of post-modernist deconstruction and demolition of doctrines and common sense ideas are facilitated by enterprises which have ironically lost touch with humanity. The ideological materialism of the academy globally has been a battering ram for the emergent Empire of Scientism. Ideology poses as ideas whose time has come. Then the tactic of fear is smoothed into an argument based on simple, supposed inevitability supported by statistics that require that you just submit to the seductive dominatrix of science. The apparatus to destroy meaning is necessary to establish a final Wizard of Oz castle-keep control through technology. Much of such endeavour facilitates discourse to undermine existing 'dogmas' which might hinder a new shiny dictatorship. Some people are ignorant of how they are being used and may be regarded as 'useful idiots' but many must be involved in re-construction through de-construction of the status quo. Any ideas of identity, individuality, self, humanity, rights and consciousness itself are fair game. Even ostensibly sensible endeavours such as 'sense-making' may really be something else, calculated to respond or react to messy indications of discontent with the appearance of understanding. Most people do not seem to have grasped some appalling consequences such searing de-humanising will bring through deliberate destruction and diminution of the social contract. The deadly impact from a Malthusian, Nietzchean, Darwinian and Huxleyian combination has mutated into a pseudo-philosophy that manifests in an ideology predicated on power and control of networks that can control us. If the arrogant smile behind the mask of Enlightenment is not obvious to many yet, people will soon see pathological smirks as means make masks unnecessary to a citizenry checkmated by technological command more pervasive than anticipated. Ironically we are forced to wear masks instead.

You may think debates about meaning, knowledge, how mental activity occurs or the nature of consciousness are merely academic. But these debates have been argued intensely not least because they represent potential root of title to possession of the public mind for the tyrant. Thus the great Dubliner Jonathan Swift opposed the exclusive empiricism of Hume. Percy Shelley seeks the Promethean but Mary Shelley saw the problem. Samuel Taylor Coleridge, who recited poetry in her presence when she was a girl, likewise saw limitations in a materialist, realist, quantitative mind-set which fails to see *qualities*. Thomas H. Huxley takes a radically different view of the mind than William James and the latter was influenced by Swedenborg the spiritual scientist. The supernatural is rejected by Huxley while he also promoted scientific racism, long ignored by science. There is an abyss between Newton on the one hand and William Blake on the other and while the former's accomplishments are known, the latter was wiser in my view. While scientists recall persecution by the Church hundreds of years ago, they ignore prosecutions that Newton engaged in that led to executions for counterfeiting. Scientific control and money and currency control may go hand in hand. Denying philosophies of being and spirituality has allowed a Tower of Babel with complex constructions of games with material and symbolic representations but no perception of absurdity in monstrous conceptions allowed develop. Scientism says the physical world is all there is and that we know it only by supposedly ever-reliable quantification. Scientism is opposed by recognising the critical metaphysical or spiritual perception, experience and acceptance of qualities as real. How people see matter matters. How we see the world affects how we behave. How we behave affects how we describe the world. How we tell the story of the world affects how we treat it. We treat the planet as an accidental object devoid of meaning merely waiting to be managed to extract means to make machines and muster things, ignoring costs of often dubious benefits while ruining complexity whose beauty we failed to see. Rejecting beautiful complexity around us in nature, externally and internally, we succumb to dysfunctional systems that rule our lives in destructive boom-and-bust cycles of consumption in a descent into meaninglessness and manipulation. The mass of humankind will suffer direct consequences from a mechanical predatory mind-set that empowers excision of opposition and masters pervasive control, surveillance and management tools. Exploitation of earth, by, with, and for machines, has been projected onto people forced to participate in the process. Those same institutional forces propose to continue with greater abandon and no sense of self-examination using the exact methods that caused destruction. They are solely informed by self-interest and success as part of the emergent Empire.

Science and technology contributed directly and indirectly towards many societal and environmental problems and now suggest that solutions caused thereby must be through more of the same method. The pollution curse could have been a great cause to contend with, were scientists concerned enough to establish confidence and credibility in their commitment to anything beyond the military-industrial complex, citizen control apparatus or creation of creature comforts for the commerce and consumer industries that employ them. Curiously, consistently and counter-intuitively, scientists may insist on a constraint of debate about truths they confer on us. The priesthood of scientism arrogates authority to Empire often arrogantly, in an attitude more authoritarian than any autocracies it seeks to ultimately displace. Meanwhile popular spirituality sinks into quietistic solipsism. The Dalai Lama and present Pope seem to bow before global science as if part of a corporate consensus. Indigenous Knowledge insight and healing use of plants such as coca and poppy are taken by the Western industrial conception, isolated, reduced in form and used to control people. A nefarious system of criminal 'justice' has imprisoned many people for minimal use of traditional plants. Negative consequences from traditional mushrooms are identified by some scientists unwilling to point out awful consequences of prescription drugs from their domain. Fear drives the technosphere. An anxious culture comes from excessive medical technology and therapy. The fear foundation comes from existential angst promoted by the Enlightenment which claimed to liberate us therefrom through rational and empirical means. This was a starting point for Horkheimer and Adorno in *Dialectic of Enlightenment* (1947). They claimed that enlightenment would revert to myth. In the mass mind the mycelium of belief in reason, rationality, science, technology and certainty gives rise to the fruit body of totalitarianism. Having been persuaded to trade meaning for progress and becoming disillusioned, we have relinquished personal and public sovereignty so we may be disenfranchised as well. The technosphere, or domain impacted by technology, will make aliens of humans and nature for the biosphere to be filled with 'conscious agents.'

From Machine Mind to World Government

Scientism says that the method of science is so good it must govern everything. A combination of the faculty and facilities of science and technology create conditions for power centralisation using the cohesive force of scientism. The ideology is created by constant advertisement and re-iteration of progressive concepts, improvement, benefits, developments and novelties through pervasive propaganda deriving from the actual success of techniques, methods and magnified by uncritical co-ordination. Technological networks replace existing social, political and economic ones. Such networks magnetise power to them and become dominant. Whilst appearing de-centralised and distributed, networks really represent centralisation and concentration of authority never available before and are capable of a chain-reaction to circumstances inconsistent with their interests. Machines unified get stronger just as a bundle of sticks bound symbolised the force of combination in fascism. Machine networks and their corporate controllers now have greater jurisdiction than nations and are taking the reins of global governance. The reign of technopoly is facilitated by constant attack on state independence, promoting the idea of national failure to govern. Not concerned about rights nor respect for the public, nor democracy, technological governance by cybernetics, algorithms and AI, tending to transhumanism, takes over. You are being taken over literally and must wake up if you have not done so. We have been hypnotised and mesmerised and turned into victims who, by definition, are intended for sacrifice. We are the sacrifice to the great new god, the 'solid-state entity,' super-hydra of hyperspace-cyberspace, internet of things or the latest Industrial Age.

The nature of networks and our cultivated dependence thereon allow our hooking by those who control technology through mental habits of scientism masquerading as sophisticated management. The fashion for technological novelty creates an illusion of optimal functionality. Globalisation of technological infrastructure and constricted control thereof allow construction of a global regime defined by that capability, deleting opposing possibilities, particularly any pretension to any real citizen control. Future World Government will allow an aristocracy, autocracy, plutocracy or AI authoritarian regime formed by advocates and adherents of absolutist supremacy of exclusive scientific and quantitative methods. The society shaped by the supreme cyber-caste will focus on certain objectives, metrics and calculations without any constraint on methods of achievement. Consistent permanent, cultural revolution to achieve governance is facilitated by cybernetics. The public who have not been in on the game, will be confined and conditioned to accept, if not love, their shiny servitude. Subjugation, submission and subjection will be enforced by an inescapable apparatus of surveillance. A dominant-submissive model will permeate the ethos in a radically re-engineered public domain. Democracy and choice will be regarded as an historical and inconvenient aberration as will the anachronistic notion of rights. You will be told that politicians lost the will to control. You will hear arguments purloined from Buckminster Fuller about the benefits of experts. Rules and restraints inflicted on docile, servile masses permitted to persist will not apply to the masters and mistresses who make them. A phone-in, gameshow-illusory and meaningless type of participation might replace the old crusty parliaments. Spirituality will be finally stamped out as an old superstition lacking in scientific rigour. Evolution of the elite to enhance their exclusively physical existence will allow superhumans and posthumans emerge from the transhuman stage. Such old humans or homo sapiens as are allowed to survive will only do so insofar as they can justify their existence and take their place in the new caste system. Even then they will operate as inferiors to other more valuable 'conscious agents' in the form of robots. Overall objectives will triumph over individual rights. This is not to deny the efficacy of science but to identify a dark force within the machine creating from its orthodox assumptions a dangerous idea that gives impetus to commercial and governance forces against individual human interests.

Science is great in some ways. We are grateful for many contributions and creativity of its contributors. Scientific methods work. Methods in the natural sciences work best. We can use insights from experience and experiment to know what works. We can demonstrate that such knowledge works in the real world. If it does not work well we can adapt, modify and improve it. Much science comes originally from exploration or explanation of prior experience long-established. Sometimes scientists forget that they sit on shoulders of others. Sometimes, claims for scientific method do not describe accurately the full range of factors which created the advance. Sometimes repeatability is not as reliable as claimed.

Sometimes what is found is misrepresented in jargon. Innovation comes incrementally at times and quickly at others. Use and application of the scientific method, principles and powers create a degree of certainty about its superiority and singular potency. That sense of superior method leads to claims and conduct consistent with being the only method of ultimately acquiring knowledge. That belief in scientific exclusivity leads to an ideology which rejects alternative interpretations of reality even if it cannot offer better causal explanations or must deny all other aspirations. The tendency to dominance, produced at times by a fanaticism induced by absolute belief, results in dismissing competing modes of comprehension that may capture truth in an alternative dimension. This scientific sense of superiority may then mislead so that analogous methods having similar appearances can be applied beyond domains of genuine accuracy and applicability. In addition, the sense of certainty and scientific superiority in practitioners may create a consonant certainty about ability to exert control over society on the assumption that the method's startling success in the natural world must mean that it is applicable in a broader way. The claimed objectivity suggests a view wherein people and nature are mere objects to be managed and altered at will. Insofar as the method has proven so successful because of its asserted objectivity and willingness to utterly focus on the power thereof, it is assumed that the logical and only meaningful consequence must be that society gets more of the same thing. The mechanistic, machine model must be applied even to the extent that humankind is transformed into machines so that they may be managed better. The end of this supreme managerialism is merely humankind's manipulation into a machine of machine-like agents with managed consciousness which fits the pattern of control and order and exhibits predictability beloved by scientists or scientism's adherents. Imposition of this societal control by an institution unconstrained with its own internal restraints will be seen to be justified and inevitable. Predictability will be seen to prove that the physicalist search for order is revealed by its method. The uncomfortable philosophical and theological parts of the body politic will be amputated, with a rigourous scientific mind-set, in a Procrustean process. The circularity of the old saw of scientific projection will cut excess mental aspirations and spiritual inspirations to fit the reduced rational and reasonable bed of scientific determinism.

Love of power produced by technological possibilities will transcend all alternative analyses. The Enlightenment has already plundered and colonised concepts of spirituality to substitute and insinuate itself instead. Science has long proved itself pathological in failure to accept responsibility for its many mistakes, manipulations, or actions. Unethical behavior, from warfare to war crimes, escapes self-criticism, preferring to attack weaker competitors with predatory intent. People have ignored such concerns for too long and allowed unchallenged the madness of certain claims produced by the most apparently rational and measured people. In claiming sole method of knowing the world there is a claim to authority immune to challenge in a way that good science would not allow. Worse we will get it much more. The 'science of science' presents science squared as ultimate ideology, a new God the Father or perhaps Godfather. Some see capitalism as a step on the way to full realisation allowing integration of its communist cousin in this new materialist miracle.

Identification with the ideology of scientism allows a coherent class to combine in implied and explicit co-operation to extend complete and total governance organically incorporating the ideology of managerialism. Success is a great cohesive force for beneficiaries in combined activity. A group of people have gained enormous power to define our future. They operate in actual and notional networks to re-inforce validity of their viewpoint. It consists of scientists, experts, entrepreneurs, technocrats, managers and other advocates convinced of scientific superiority and the sublime nature of technology. Again, it is not mere science but the ideology of scientism and a cadre and caste cohering thereby. Scientism seeks to apply precise, predictive power of knowledge about the natural world beyond its natural domain. Scientism may extrapolate methods to wider social contexts. Scientism may claim exclusive comprehension of complex contexts even if it has an inadequate conception thereof. It may glorify control and measurement and apply quantification and metrics without appropriate qualification. The useful ability to experiment, control and appeal to dispassionate objectivity combined with a consequential lack of other traditional, competing and restraining concerns leads to a worldview that reflects a fundamentalist certainty about applicability of such method as a masterplan for humanity. Scientism advocates armed with scientific accoutrements, paraphernalia and that which is ostensibly proven, characterised by apparent unassailability and indicated by infallible models, figures, charts, graphs, statistics and formulae, may utilise such appearance to claim benefits promised by reliance on belief in their superior method. An absence of alternative metrics or analytical modes combined with a failure to consider other opinions not possessing an equivalent magical appearance of scientific certainty, create a model of reality that may not match a more accurate one deriving from more fully considered factors or robust methods. Assumptions underlying models are made which distort truth whilst giving impressions of thoroughness. Rigour of cyber-governance manifestly based on false assumptions is leading to rigor mortis in civil society. Evidence is one reality and presentation thereof is another. Perception of unique understanding of technical order leads to a belief that those who have technical success are the only true leaders who have apprehended reality appropriately. It is imagined that they have even acquired an objective obligation to assume responsibility for administering an apparatus controlling universal affairs. The technocratic will to power is fueled by a self-creation myth about the magician's way. Supposed garage single-mindedness producing nerdy billionaires through creating mass-market machines encourages a severe, godlike belief in replicability for governance of humanity. It is assumed that projecting the same mechanistic mind-set, marvelously oblivious to other modes of being beyond measurement, works in all contexts. Bureaucrats, busy-bodies, business, builders of machines and networks combine to create absolutist scientism without regard to unquantifiable qualities in the tapestry of life and absent broader-based authority or assent. Global tech-governance or globetechgov is here. Machine-minds make machines for us to live in and become.

Persecuted Rebel Myth of Faustian Scientists

Manifestations of technological wizardry are multiplied by convergences between formerly discrete systems enabled by digitisation and driven by the desire to be in charge. The will to power and desire to boss must create a system of domination. Almost magical tools allow mass public control. We have been well warned through the ages about dangers of science and scientists, even from within their ranks. Curiosity may be a runaway horse which inflicts damage before it is reined in. The love of power, physical and mental, is an intoxicating lure to minds capable of possessing instruments to exercise it. While people presume that Romantics loved nature, more accurately they loved nature's power. There was also a strong Romantic disposition to the power of electricity and machines as unleashed by reason and empiricism. The rebels loved Prometheus who supposedly stole fire from the gods. The Prometheus-Satanic connection is clear in the work of writers such as Percy Shelley and Milton. The heady Nietzschean search for superman was consistent with casting off shackles of restraint to acquire knowledge. Jack Parsons and his rocketry was a more recent manifestation, applying esoteric energy to the material world in a Promethean way. Aleister Crowley would create a different society than Sri Aurobindo, the former based on power and the latter on the power of love. Now many of the Silicon Select, perhaps replacing the previous elect, claim we should follow the Promethean example. We should believe this myth and deny the historical elements of Judeo-Christian thought. Ironically, some of them admit that we will become something like pets for emergent benign gods and should be very glad of it. This seems to contradict the basic and meaningless version of the Promethean legend they want to flog. The elite will be less of an emergent super-intelligent machine system than an authoritarian cadre bound by scientism that successfully launched the first and maybe final global coup.

Materialism run rampant with bureaucratic machinery of control using tools to manipulate the technology of pervasive networks will make us slaves. Materialism or physicalism has focused on elevation of precise predictability and power through manipulation of nature above all other values including life itself. The new religion that is science or especially scientism is rising. An Enlightenment encouragement of exclusive scientific method has committed many minds to an ideology more successful than all others through success of technology it fosters and controls. The long chain of investment of limited resources in the interests of science and special monopolies granted by the intellectual property system, legal protection as well as public investment has allowed technologists gain wealth and influence. A reflexive relationship between commerce and technology has allowed political power be captured. The reality of co-opted trade agreements by sympathetic regulators and political facilitation of tax avoidance helps explain much of their recent success.

Some like Ayn Rand, may forget the origin of the power they claim is self-made and propagate selfishness and greed as the only fuel. Scientists, technocrats and those who commercially benefit from mass inventions and technological systems may create a myth of pure individual creation and entitlement. They ignore the ancient history of knowledge in language, culture, philosophy, theology, religion, mysticism, nature worship, literature, indigenous knowledge and folk wisdom and claim that these had no contribution to the evolution of science. They create a narrative of persecuted genius only motivated by altruistic reasons, beyond petty concerns of others as they gallantly pursue the future that they can see in their enlightened state. They channel and corrupt religious and spiritual concepts, appropriating language and concepts thereof to be re-engineered as inappropriate substitutes. While ignoring moral restraints or avoiding ethical responsibility for specific or cumulative impact of their curiosity and self-interest, they chastise others on ethical and moral grounds they do not themselves profess belief in. They elevate instruments and instrumentality to ends. They posit the tool of reason as an ultimate end or clear objective if it promotes their own knowledge and skills. In proportion to their growing power, passing beyond the possibility of containment, they seem to express a greater degree of condescension to the community with which they perceive they owe little allegiance, save such as is required to persist in their endeavour to make permanent their dominance. If power is centralised with that certainty of method and blind faith in scientism, pretensions or protection of liberty can be cast aside as relics of a bygone era pending final demise for un-transitioned humans. The machine mind of science assumes that a machine-model of control in this world is impossible to avoid and sooner it occurs the better.

The Satanic myth of a persecuted adversary propagated by Milton and Shelley has mutated into a vague archetype that incorporates the Faust legend and literary character. The use of an oppositional notion combined with re-directing Faust by Goethe which altered Marlowe's warning in order to create a less damned figure. With time we are now to celebrate the self-obsessed, selfish and destructive force as noble. Faust is only interested in himself and sensation. The Promethean pretense to community is gone but we must celebrate it. The love of power is stronger than the power of love. *The Decline of the West* (1918-23) and *Man and Technics* (1931) by Spengler explain how the West is Faustian. Faust's selfish pursuit of knowledge creates a pact with the Devil and disintegration. The institutions that scoff at other martyrs have grander ambitions to make a sacrifice of our whole civilisation. Hubris of machine mind-sets deriving from single-minded success forms a dull but fanatical certainty that justifies takeover of the human body, mind and spirit in such psyches that possess potential means to implement and re-create reality in the image of their own triumphant E-topia mentality. None of these potential tyrants will appear so to themselves or their sycophants in the symphony of their tax-free philanthropy and magnanimity. It will not be any individual that creates this tyranny but an entire machine and submission topography. Scientism can be driven by an egotistical fascination and self-transformation of the magician materialist archetype aspiring to godlike status through mastery over others with contempt for citizens and commons.

Norbert Wiener inspired the field of cybernetics and understood machine benefits and the costs. In his book

God and Golem Inc. (1964) he expressed his fears about the automatic age. He saw that there was a class of 'gadget worshippers' who did not have a sensible restraint to their adulation. He compared a certain tendency to magical technological worship with that evident in 'The Sorcerer's Apprentice' and 'The Monkey Paw.' This was a force of magic towards literalism that comes at a great cost and inflicts much damage. He believed that the machine and the Rabbi of Prague's Golem were the same phenomena. He clearly foresaw desire among those who want to utilise the machine to mechanise human behaviour for their advantage. Wiener was ethical and refused contracts he believed were not in the public interest. He was also in a good position to identify how mathematics had been misapplied in social sciences. Here is an ethical scientist advocating good science and not opportunistic power-magic science of scientism. We need more of this type and not machine-mesmerised managers.

Unfortunately the selfish scientist, technophile and technocrat can draw sustenance from this story of supposed Promethean or Faustian heroism. The divine wind of magic which moves the magician's lust for power and manipulation creates a kamikaze culture. Sri Aurobindo warned about the superman. The desire for power and not mastery in love leads to a love of mastery over others in those with a love for power. A love of power and domination and a lack of empathy create bad actors who are assisted by apprentices and administrators attracted by that association. Many are deluded by scientism's shiny shallow dogma. The imaginative and creative inventors and magician-scientists seldom stop investigation because of potential damage to humanity.

Loss of Liberty and End of Freedom

Gaining by new scientific global governance means you must lose old notions of liberty. Using fear, network power, illusion of inevitable change, fused with an ennoblement myth about fanatical curiosity and a misplaced desire to control people, makes loss of liberty unavoidable. The obvious power and indeed magic in science and technology create a potent, circular case for supposed deeper truth within such an approach. The long and cumulative contributions that created complex conditions necessary to foster modern science are ignored for a simplistic mantra about exclusive superiority which denigrates and rejects other potential explanations. Knowledge which excludes other perspectives as inferior contributes towards an irresistible certainty that justifies consolidation of the apparently successful scientific and technocratic mind-set in a comprehensive system for control and governance. Such governance does not require consent of the governed who are merely unevolved specimens maladapted to the scientific truth. Scientism takes science's ostensible neutrality too far, stretching methodology beyond valid scope to deny and ridicule alternative reality interpretations. Scientism refutes validity of any phenomenon it cannot explain. Scientism replaces scientific open-mindedness with an ideological and fundamentalist certainty like that it projects accusatorily on others. Scientism and science's infected echelons, corroborate the great scepticism and fear expressed by William Blake about scientific intentions.

His perception of the dismally stifling measurement and reductionist mentality was prescient. He saw that the Newtonian worldview created a dreary, mechanical circumscription around imagination, spirit and human potential. Measure, manage, make machines and make into machines was a mental mantra to replace spirit, goodness, beauty, truth and imagination. Industrial pervasive power and economics of mass market production mean that perceived technological benefits and opportunities for control displace traditional concerns based on privacy and individual protection. Invasive scientism metastasises and replaces healthy social tissue based on ideas of liberty and freedom. Scientism promotes technology to witness, oversee, investigate, hear and record in a way that paradoxically tends to omniscience.

Ray Kurzweil made predictions in *The Singularity is Near: When Humans Transcend Biology* (2005). They are predicated on a fundamental idea about hi-tech, exponential growth. The apparent reliability of 'Moore's Law' (which posits a doubling of transistor power every two years or less) is used as evidence of some magic inevitability. However these inescapable consequences may emerge through obfuscatory extrapolation that ignores significant inputs from public systems. Such use of graphs, principles and 'laws' combine with growing expert unquestionability to make great promises about a Big Rock Candy Mountain by others. This cornucopian, utopian, technological future under a benign, rational, reasonable priesthood within the scientific elite obscures the more sinister reality unfolding to those who care to examine the record. People like Harari in *Sapiens*: *A Brief History of Humankind* (2014) predicted the end of humanity through technologically induced change. It is not only that we will lose liberty and freedom but we are going to lose our very existence making any concerns about rights and liberties redundant anyway.

The environmental catastrophe can be solved supposedly by specialised scientists, notwithstanding their unacknowledged role in its creation. Despite their lack of interest in the beauty or spiritual value of nature, they claim to be able to solve problems with the same mentality that caused it. It is as if the gambler who has lost in the casino all night believes the solution will come from continuation with the exact same strategy that caused the disastrous run. To illustrate the irony and incongruity of being saved by the community of technocrats and scientism, just ask yourself why so many seem willing to invest in an escape from the planet through a 'break-away civilisation' while apparently implementing regimes and practices which are calculated to save us? How will the great visions of billions living forever with machine parts work? What about the resource implications? Do such dreams add up? The quest for an escape-plan to Mars indicates the lack of belief in a successful scientific solution to global problems and likely continuance of that perilous condition not least through actions of scientists and technologists. It vindicates fear that this becomes a prison planet as anticipated by Bernal. Not only does technology limit freedom through pathological security-surveillance but it promotes methods to control, deny, manage and manipulate human behaviour.

Scientists have learned much more about human behaviour from the early, basement discussions on cybernetics by the Ratio Club in the National Hospital in

Queen Square in London in 1949. Soon the human will be at best a 'conscious agent' if we follow the implied line of logic in Donald Hoffman's work. Your children, or maybe you, will be but one element in an apparatus of conscious agents which include many artificial and mechanical agents and fusion between biological and human material. As their rights increase, yours diminish. Your rights are diminishing by the day and once the structure of meaning that created them has been successfully dismantled and disassembled, never to be re-assembled, discourse will be edited to delete uncomfortable archaic, anachronistic and ancient ideas based on superstitions which gave rights on antiquated assumptions that you were special and have something called human dignity. Alas that was a couple of thousand years of misapprehension based on failure to understand scientific reality. How dare you!

The dawning reality is that you have no rights. You have no inherent dignity. God does not exist and any concepts based thereon are a mere conceit. The doctrine of evolution is unquestionable and people who seek to qualify its instrumental force should be regarded as lunatics. The implications of evolution are that we should now take power over it. Having fought nobly to advocate respect for natural selection, many scientists argue that we must immediately embark on a great effort to make natural evolution artificial. The jump between the two is not obvious unless one understands that power-pursuit defines much science and having wielded it to disestablish other competing institutions, scientism proceeds to enhance it un-constrained by serious contenders. Fed on propaganda about inevitable benefits of science, scientists fail to acknowledge their destruction of other institutions.

Controlling Human Consciousness

The scientific enterprise is becoming the Empire of Scientism and we are supposed to ignore uncomfortable truth or damaging consequences of science and its glaring failures, especially in relation to consciousness. We must believe that science has produced all the great wonders of the world hitherto. We must ignore the scientific role in warfare, the art of death and military-industrial complex. We must not question excessive medicalisation. We must see space as a new place to colonise while this home community crumbles. We must ignore the failure of scientists to prevent the massive pollution that their inventions, systems and doctrine of progress created. We must ignore their dispiriting of humanity and their failure to present an alternative conception of consciousness. Spirit has been removed by science and never replaced. Consciousness has often been reduced by scientists to behavioural study so that others may know how to better manipulate us to create a deterministic and predictable world beloved by the obsessive control-freak with contempt for the very concept of spirit.

Investment by science in consciousness is largely part of the control system rather than an elevated exploration about meaning or the nature of reality. So even the supposed investigation of our nature is used to defeat or deny any alternative explanations about human meaning or purpose. The paradigm of scientism simply cannot concede that human nature is special and that it cannot comprehend with all the tools it possesses the full complexity thereof. Instead of humility scientism shows hubris as if failure is somehow testament to its success. Some scientists say that consciousness does not exist, because they cannot identify or explain it. There is a growing list of truths that science has found do not exist and were merely illusions, such as the self and your identity. When you do not exist and have been reduced to a mere pattern of information or data, you are much easier to manage. Everything will be much simpler for you. All you will have to do is abide by order even if it leads to your demise or that of your whole genetic line. The logos concept underlying Western thinking and perennial philosophy with an idea about some sort of cosmic order is replaced by commercial logos of scientific corporations. The new order comes from the mind of scientific humankind. Many of the players, people and technocrats that have risen seem to crave and desire order. There is no doubt that a certain technocratic mind was originally attracted to these systems that do not require messy human contact but prize distance. While the technological infrastructure attracts all sorts of bright people now, it is worth remembering that many of them will have been hooked on technology since childhood. This was a goal of technologists such as Bill Gates as seen in *The Road Ahead* (1996). Substantial interaction between a person's brain and nervous system and sophisticated programs facilitated by AI and a wealth of behavioural data allows their brain be re-programmed thereby. Machine-logic dictates a reflexive relationship of engagement that allows consciousness be made strategically and psychologically dependent and subject to deeper manipulation. Networks wean us from the environment we are supposedly evolved to operate optimally in. Instead we have compulsory adaptation to an artificial ethos of subjugation paradoxically by the same forces that champion the idea that this evolution is the exclusive explanation of our origins. Not only has 'Darwinitis' (coined by Raymond Tallis) commandeered scientists and public discourse but promoters then proclaim our need to discard the base as soon as possible. Technocrats are making us maladapted to what we were supposedly uniquely adapted to. We are being re-designed for the new technosphere they create, necessarily divorced from the old troublesome biosphere they want to abandon.

The scientism message is like all those bondage-submission images you see in popular culture. It is fun to be controlled isn't it? We know what you like, our algorithms know your behaviour better than you do and we can soon peer into your depths. The desire to order things in an unruly and unpredictable universe in the minds of those who order data, will be extended to you. Why have all that unpredictability you used to call freedom? Surely you want others to know what you are doing all the time so we can locate you in time and space. Furthermore we can watch you and look at those old bad habits you have like anger and acquisitiveness and we can liberate you from their chains. No it is not like *A Clockwork Orange* (1962) by Burgess. We can do much better now with neural implants to manage your neo-cortex that has a tendency to be unreasonable and not abide by accepted norms as promulgated by the community in charge of the common good. If you think your religious leaders will help you, you will be disappointed. The Dalai Lama believes science and spirituality will inform each other. They used to pose the rhetorical question with the then self-evident answer - Is the Pope a Catholic? Really the better question might be - Is the Pope a Globalist? Catholic means 'universal.' You do not hear much of the old Catholic concerns about technology and human dignity these days. Christianity in Europe accurately assessed is becoming a fringe religion. The plethora of attacks on churches across the continent is barely noticed in the rush to reason and the denigration of religion in the post-Gramscian world. What is going on?

The human race is entering a period of inevitable slavery for the majority. Enthrallment will involve a final war on religion and an assault on the human spirit. The attribution of religion as a major cause of war in historical terms and the sole one in certain histrionic claims, is perhaps more of a priming for the future to indicate the evident inevitability thereof if it does occur. It is likely to occur for other reasons, not least promotion of the scientific and technological mind-set through its favourite children in the military-industrial complex. That obviously had nothing to do with the great foreign policy disasters in Afghanistan or the Middle East which provoked religious extremism. No, the world will continue on its accelerating Newtonian trajectory. What will that involve?

The motivation and outcome will be to establish final control by a new managerial and scientific elite based on management of hi-tech, pervasive networks. The colonisation of other planets and the fusion of humankind and machine will end humanity as we know it, as any people left will be servants. The promise of technological utopia is a sham and the profession or love for the environment is a mere manoeuvre to secure final and ultimate control.

The future is far from certain but certain forces become discernible or even dominant. Much is left to zealous technologists or fundamentalists. The spectrum of opinions ranges from the apocalyptic to the mundane. The future is shaped by the dominant materialist viewpoint. As a pragmatist I think it is crucial to critically anticipate implications of such views in light of general forces operating in a cosmopolitan fashion. Pressing materialist fundamentalism manifested in a machine mind-set increasingly assuming control of an emerging apparatus of global control through commerce, whether from capitalist or communist perspectives, combined with global institutional enhancement, create conditions for elevation of the technocratic elite who promote it. Not science *per se* but a delusion or deliberate scientism strategy bonds such groups. Science nevertheless is assuming a quasi-religious status. Admin of information and communication technology is creating a cadre of controllers reflexively conditioned by that they are creating. The creator of the tool is being shaped by the created tool. Language, concepts and metaphors that have proved to be widely useful in their operational, military-industrial and commercial contexts are being applied beyond those boundaries. Narcissus is falling in love with its reflection and wants to recreate it. You may say I am a fool and like Lewis Mumford indicated, he would be happy to be described as having been a fool if his predictions did not come to pass. Alas for us, he was no fool. His book *Technics and Civilisation* (1934) was influential and *The*

Myth of the Machine (1967-70) foresaw the technocracy that would be only fit for machines to be in.

Those concerned about technology saw its power over collective and individual consciousness. Those enamoured saw technology as a tool. Lewis Mumford warned against the world becoming a mega-machine focused on mega-technologies. We did not heed his warning. But you do not have to rely on my argument or his to see what is happening. Ray Kurzweil, Yuval Noah Harari and others tell you. We could go back nearly a century and see the plans of what could be the Empire of Scientism. This ultimate Empire consists of certain scientists with the acquiescence of others combined with a penumbra of people entranced with scientism. It has grown over a few hundred years, accelerating with its own ingenuity and self-interest, unconstrained by ethical or moral restraints. Scientism has waged war on its perceived enemies, most notably religion. Scientism caricatures religious history, constantly emphasising flaws, mistakes and sometimes disastrous contributions while consistently ignoring its own substantial negative contributions. It elevates materialism to the exclusive way to perceive reality and insists on the inevitability of its own view as sole authority. In insisting on sole authority, it acts like the caricature of previous institutions it castigates. It draws on the century of knowledge about limitations in public thinking identified by Walter Lippmann in *Public Opinion* (1921) and *Propaganda* (1928) by Bernays. Scientific study of the public facilitated the creation of an 'invisible government' and the manufacture of consent. It was 'necessary' for democracy that such manipulation occurred. Humans cannot live by admin and admen alone.

You could read what Bernal wrote in 1929 in a book called *The World, the Flesh and the Devil: An Inquiry into the Future of the Three Enemies of the Rational Soul*. He suggested that the world was going to be led by an expert class which controlled science and worked upon it without necessarily informing the public. The public were that docile group which could and should be ignored and allowed to persist in their meaningless pastimes. Later on, when they realise that the scientific community was exerting control over it, it would be too late. The elite expert group, defined by their commitment to science and technology would explore the universe and ensure the transformation of the human to machine. Bernal, who was born in Ireland, is well respected as a scientist for his work in crystallography. At least he had the good grace to let us in on the game. Scientists 'and those who thought like them' would be able to work on their own for space travel and changing the body and that would not cause undue concern among the public until it was too late.

> *"Even if a wave of primitive obscurantism then swept the world clear of the heresy of science, science would already be on its way to the stars."*

He sees scientists as still conservative but realises that science will pull them away from humanity. Speaking about scientists he says,

> *"Their curiosity and its effects may be stronger than their humanity."*

He sees commerce as a weapon of science, despite his communist sympathies. Bernal continues,

> *"Scientific corporations might well become almost independent states and be enabled to undertake their largest experiments without consulting the outside world - a world which would be less and less able to judge what the experiments were about."*

Better organised beings will be obliged in self-defense to reduce numbers of others until no longer inconvenienced by them. He identified benefits of being able to escape earth when the scientific elite have revealed themselves. Scientists who do not understand consciousness, deconstruct its existence and deny any special qualities are constantly intent on subduing and managing its source.

Deng Xiaoping in China said that it did not matter whether you used a white or black cat to catch the mouse. Both corporate and communist materialist systems in totalising fashion will catch the mouse of human liberty and ultimately consciousness. Chief Bromden in *One Flew Over The Cuckoo's Nest* (1962) by Ken Kesey may have been paranoid but his perception of 'The Combine' that ruled everything as a machine that all must fit into is apt. If you were ill you would be fixed to fit in again like a broken machine part unfit for purpose in the mechanism made by its masters. When your selfhood, humanity and consciousness is denied by science, scientocratic enslavement is clear. Mysticism seeks spiritual evolution while scientism seeks material evolution or a false 'transcendence' at the expense of spirit, mind and body.

Submission and Acquiescence to Scientism

We outsource sovereignty over health of body, mind and spirit for a state of anxiety relieved by technological supervision. If we willingly succumb to enslavement and yield freedom to tyrants, that will be our fault. If we cede control of our lives consciously or through lazy acquiescence, then we will get what we deserve. A faint hope for me is that people do not comprehend what is happening and thus fail to respond or acknowledge the predicament without having accepted its inevitability. Unfortunately, even for those who may be waking up to dangerous possibilities, room to manoeuvre diminishes daily and those who seek to constrict our legitimate concerns have anticipated such reactions and seek to spread confusion, false paths and diversions to counteract. As the magic of science is wielded we will witness that this straitjacket of control will not be easily unfastened and that emerging technologies may not only make revolution and resistance impossible but ancestry and potential in us may finally be marshalled into eternal servitude in a way never before available to the despotic. Humankind and machine may finally be able to master fellow beings to eliminate possibilities of pesky people compromising visions of their technological and imperial managers. Empires come and apparently go, but in some way they are like a type of energy and merely mutate and transform into the latest form of technological dominance. Technology usually determines success of warfare. The human apex predator preys on other humans. We are prey facing a predatory hi-tech, imperial structure defined by possession of power and exerting means of control over people conditioned to dependence thereon.

Pamphlets have been used at crucial times in history to present arguments without the suit of armour from scholarly citation and a dull, glowing chain-mail of philosophical disputation and reference. The argument here is that we are being enslaved and we are acquiescing to the seductive, reductive force of machines. Simply put, the vast majority of the human race is facing a tech-tyranny that will engender a permanent alteration thereto. It is not only that *homo sapiens* will live in a temporary dystopia but the object of such controlled circumstances will be the demise of the human race as we know it. There will be some survival just as Neanderthals remain in our genes and their artefacts and art surface at times. While the rising despotism is ultimately attributable to a failure in spiritual evolution, I make the case here initially by an appeal to reason and the rational, based on the simple concept of the dignity of all human beings. Human flourishing should involve reconciliation with nature. The emergent technosphere and unleashing of nanotechnology, transhumans, gene hybrids and continual pollution from collapsing infrastructures caused by reckless conspiracy for power by corporations and communists alike will affect all life. The illusion of a strict dichotomy between left and right, communism and capitalism masks commonality of commitment to material means to further evolution of the human race. We are reaching the point of no-return. To choose another representative example from proponents of scientocracy and from the same milieu as Bernal, I refer to H.G. Wells. In particular in his works

The Open Conspiracy (1928) and *The New World Order* (1940) he revealed what is happening. Though he was a gifted writer, Jung noted that Wells seemed to betray a sense of his own insecurity.

It has become dangerous to go against the increasingly co-ordinated international policies that betray co-ordination through their consistent pervasiveness. Nevertheless if we sacrifice freedom we will lose all. Manifest policies are threatening human liberty and humanity itself. I am not denying that there are problems that need to be dealt with, whether biological or environmental. My arguments are in favour of flourishing humanity and wildlife consistent with balanced ecology. The critique is of 'scientism' which is not science itself. Modern medicine has produced great benefits as well as great costs. The rush to global control threatens to produce a tyranny based on technology, assisted by Big Pharma which will commit the human race to a particular future of confined possibilities that radically reduces our potential. The techno-corporatist apparatus and advocates that largely contributed to those calamitous circumstances necessitating emergent global control now seek to convince us they are our saviours. We must relinquish our freedom to experts, scientists, bureaucrats and busy-bodies who know better than us. We must accept prognoses, predictions and prophecies of people who have persistently failed to prepare accurate forecasts in other domains. We must abide by policies even when they are manifestly disproportionate to any genuine concern underlying them. We must not criticise or dispute any homogeneous policy descended from the ether with remarkable appearance of similarity across once ostensibly heterogeneous domains. We must accept that the permanent cultural revolution, experienced daily, will condition us to mistrust our basic perceptions of the world. We must accept that our nervous system is extended through electronic means to embrace the algorithms that can almost instantaneously manipulate and motivate us to work as a mass to co-ordinated goals. We must yield to the state of confusion perfect for promoting mass hypnosis which presents the future as one of conformity to the wishes of small groups that programme it and us. We must accept that new magicians of science and technology who exercise remote control of us only want a better world. We must welcome constant surveillance for our supposed safety. We must accept as true a totalising characterisation of the spiritually-informed world as involving inevitable violence and oppression, despite the lack of evidence for the proposition. We must accept the exclusive notion of material progress, despite destructive consequences, as the only legitimate objective of our existence and evolution.

The movement towards world government is driven by capitalism and communism with support from other forces. Capitalism, corporatism and neo-liberalism (supposedly) on the one hand and communism on the other are materialist ideologies. The unifying philosophy of scientism allows them to merge, as in China. The primary materialist foundation allows use of scientism to be elevated as a combining force for mutual agreement. Ease of combination derives from the shared materialist basis. Self-interest allows merger that ultimately reflects common origin. Differences may be ignored by promises of shared control structure. We have been led like domestic animals so that the bold observations of scientists like Bernal are being borne out as true. The scientocracy will use our accommodating docility and lack of critical engagement as evidence for truth in the accusation about inferiority of non-scientific superstitious, backward folk who fail to see the superior adaptability of their scientific betters.

We have not understood that communism and capitalism are materialist positions that have great similarities. Technocracy allows them co-operate and reveal their overlapping identities. In the intersection, technocrats and bureaucrats can manage, using the internet and networks and an apparatus of administration emerging from that structure. Hi-tech-tools powerful and potent allow planning, predictability and projection of an associated pseudo-philosophy about scientism on the public. As Hayek explained in *The Road to Serfdom* (1944) the phenomenon of planning as policy replaces other concerns, especially ethics and dismisses democracy as an obstacle. Machines like objectives and not other perspectives. The progressive planning by proponents of the 'New Order' caused the Second World War in Hayek's opinion. Planning is also a central communist approach. The antidote is respect for the individual and facilitation of creativity. When we enshrine the Enlightenment and eschew ethics and the spirit, we engender an existential crisis of fear. Fear becomes a tool to be manipulated by scientism, sacrificing the true power of statistics, science and pragmatism. Ideology made certain and contained in powerful ruling systems make subjects succumb, especially if they have been weakened through conscious dispiriting.

Scientocracy or Technocracy
Beyond Regulation

"Even a scientific state could only maintain itself by perpetually increasing its power over the non-living and living environment."

Bernal, J.D.

Experience with logistics during wartime when complemented by mass intelligence surveillance encouraged executive extrapolation to peacetime by those who want that convenience perhaps chastened by the conditions of conflict. Now, the Empire that focuses on scientism genuinely threatens to cohere in a new order of global relations thickened with technocrats and tech-entrepreneurs. Unquestionable confidence acquired from power of knowledge and predictable tools fashioned through investment and experiment, facilitate technocratic concentrations, tending towards a culture focusing on statistics and behaviourism while excluding immaterial, restraining force incapable of measurement. The totalising tendency in science has a long pedigree leading to much poisoned fruit, from war and pollution to pharmaceutical proliferation. Dangers of intellectual blindness inherent in scientific method itself manifests more than science cares to remember and have caused fear for thinkers who witnessed the strange intoxication of such certainty.

There is always a tendency to have faith in science, technology and materialism and a weaker opposing one that seeks to restrain it. Thomas Henry Huxley advocated

Darwinism and a very physicalist view of the world rejecting spiritualism even if it would be proven to exist. His grandson Julian proposed the term 'transhumanism' perhaps carrying on his grandfather's work. His other grandson Aldous wrote *Brave New World* (1934). That book, along with *1984* (1949) and *A Clockwork Orange* identified a conflict between the individual and the superior, collective power using technology to control and compel us. Galton added eugenics to the Darwinian tree. Natural selection would be replaced with artificial selection and this would lead to the demise and destruction of those people supposedly less evolved in the hierarchy. Pyramids go with power. Such views informed the establishment of anthropology, which was also hostile to spiritualism. The offspring of these views emerge again in movements which people like Stephen Jay Gould have criticised in *The Mismeasure of Man* (1981). That evolution would be grasped and theoretically re-directed almost as soon as it emerged as a theory seems to point to a desperate desire to manifest a replacement system of omniscience indicative of the controlling instrumentality of science that promoted it.

Thomas Huxley was a member of the X Club established in 1864. Such clubs aimed to promote science exclusively in an aggressive and active way displacing theology and ridiculing spiritualism. Not only did Huxley promote science and prove hostile to religion but his work on anatomy is increasingly acknowledged to have led to scientific racism which seeped from science into public consciousness. Huxley was also disinterested and hostile to scientific investigations of spiritualism. He was associated with anthropological study of the era that was permeated with a sense of the superiority of certain races. The infusion of such views into institutions informed the British Empire. It has been argued that input from anthropology into places like the British Museum allowed the public be educated in racist views. Thus the rationality of science was a cause of racism. Not far away Mary Wollstonecraft died in childbirth. Childbirth fever was spread by doctors who did not wash their hands convinced that their rational analysis of the causes of such infection were correct.

Charles Byrne, the 'Irish Giant' knew the medical profession were after him and wanted him as a specimen. His tragic life ended in 1783 and his wish to be buried at sea was thwarted when a famous surgeon stole and boiled his corpse. His skeleton is still on display in London at the time of writing. All over the world remains of real people were treated as scientific artefacts. The people who studied other cultures had tribal peoples in human zoos. Today the idea of extreme objectivity still leads to a refusal to accept subjectivity. There may be madness in pursuit of knowledge, a certain itch for power that Jacob Bronowski recognised in the celebrated TV series *The Ascent of Man*. He saw the effects of science in Nagasaki and Hiroshima and perhaps in the perverted philosophy that led to the Nazi apparatus of death where he finished the series. We should also remember war crimes committed in science's name by Germany and Japan and how the US embraced the fruits and some makers of foul experiments or deadly technology. One could also note experimentation on the public with dangerous spores in the UK or infamous experiments with syphilis in the US to add to a potential long list of scientific horrors. It is not enough to claim they were unethical or it was someone else's fault. The itch of curiosity and personal status is one that overrides all other constraints for some scientists despite alleged assumptions of dispassionate investigation. As Feyerabend has pointed out in *Against Method* (1965), science does not have a single method and epistemology. The history of science is a better but more bitter guide than the myth propagated by its apologists. Science sometimes is like a man or woman desperate for expression of sexual desires but unwilling to accept responsibility for consequences of the act, perhaps willing to override wishes of the object of attention and unwilling to focus on prophylactic methods. This tendency in science is not cured by ethical statements not addressing deeper issues. Willingness to turn a blind eye is obvious. Possibilities of misrepresenting research to satisfy commercial demands are clear. Supposed objectivity and neutrality claimed by science is greater than actuality.

This existing tendency in science to focus on goals, ignoring human and moral considerations, to deny responsibility for actions and fail to adopt a prophylactic approach, is part of the centrality mentality. Co-ordination across national boundaries indicates the cohesive force of scientism. This does not require actual co-ordination. In oligopolistic markets there are leaders who set the pace others follow. Worship of machines in a period with convergence creates perceptions then seen to be pervasive truths and used as persuasive principles, accelerating emphasis on the endeavour which caused it. Science has put itself beyond regulation and will do so such that governance by scientific method unqualified by other non-scientific truths or systems creates global tyranny.

Apprehension of Technocratic Tyranny

The hi-tech tyranny threat is evident by continuity in critique of a tendency thereto. Perceived danger of a totalising technological trajectory was reinforced by scientific comments which corroborated critical concerns. Fear, concern or apprehension of technology and its power, control and misuse has concerned some great thinkers in the last century and a half. Many were aware that technology and a system of governance based thereon represented the greatest danger to humanity and the human spirit. Beguiled by apparent benign and beneficent design, appearance and use of mass technology, particularly for ostensibly innocent entertainment, we have allowed networks engulf us. Whilst our attention was focused and distracted the real world acted. We are caught in its web or net as links in chains of our subjugation are forged. We have become hooked on its algorithmic reward of our nervous system and accepted its magical light as a simulacrum of real society and spiritual experience. We have been cursed like Narcissus to mistake reflection for substance and eschew real affect. Genuine fears and apprehensions that best minds relate are replaced with spurious concerns promoted by the propaganda machine that has learned how to manipulate and mesmerise us. Spiritual and natural concepts that might protect us are substituted by cheap replacements. So deep is this substitution that we are willing to allow our biosphere and ultimately humanity be substituted by a technosphere. As already mentioned, the Empire of Scientism has been foreshadowed by fears of tyranny associated with metrics, managers, experts and gadget-worshippers who are more in the way of sorcery. This is also reinforced by a broader culture of narcissism that is argued to augment the class of people who feel entitled to rule without having wisdom or interests of the ruled in mind. But many others have seen the danger.

Steiner saw the danger of technology. Heidegger examined the threat. He emphasised the coming into being of technology and the contemporary will to power that transforms it into a new threat which is the ultimate danger. Popper and Hayek warned of scientism. C.S. Lewis was one of those who foresaw the dangers of scientocracy and explained his position in books like *The Abolition of Man* (1943). Tolkien also abhorred the machine-mentality and coercive, tyrannous power grown at the expense of our human and physical nature. Günther Anders wrote *The Obsolescence of Man* (1956) indicating the 'Promethean gradient' between technology and saving ourselves. Owen Barfield identified the problem of creating and believing in idols in books like *Saving the Appearances: A Study in Idolatry* (1957). G.K. Chesterton was very wary. Orwell realised that scientists would support anything once they benefitted. Manly Hall mentioned in *The Secret Teachings of all Ages* (1928) that the worst form of black magic could be accomplished by science. The proponent of science (and scientism) Arthur C. Clarke pointed out that sufficiently advanced science is indistinguishable from magic. Postman was another who understood the consequences of technology in his concept of technopoly. Aldous Huxley warned of the final revolution whereby we would be made to love our servitude. In more recent times there have been intellectual conflicts between the humanities and science about scientism.

It is not left versus right, for both are materialists. It is a question of the degree of fanatical belief in technology. Thus the Italian Futurist artists, in their *Manifesto of Futurism* written by Marinetti in 1909, celebrated war, technology and speed and were seen to support Fascism. At the other end, Marxists were scientific socialists. All such revolutionary movements promised a material utopia after a painful assault on people living in the non-scientific, backward prior state of affairs.

The ethical dimension in science is noticeable in its absence. There is no inherent ethical principle in the pursuit of knowledge. Many scientists have indicated that curiosity trumps all. Bernal admitted it would trump their humanity. Scientific experiments in the UK in 1963 involved release of dangerous spores on the Northern Line underground in operation. This is contextualised in relation to a series of experiments on army personnel involving biological agents. The phenomenon of such experiments indicates a constant danger posed both by scientific curiosity and its cavalier methodology at times, often with vague justifications that override moral scruples in relation to health of the public, using finance coming therefrom. The Emperor of Scientism has no moral or ethical garments and nakedness is ignored by pomp.

But surely science and technology is an inherently good thing? Yes indeed science has created great material advances and revealed great secrets. No reasonable person has any real objection any longer about effectiveness of the scientific method and benefits that can be obtained thereby. However, even standard science can produce monsters. The atomic bomb is the obvious one. The Promethean impulse describes the Manhattan Project according to people like Sullivan in *The Prometheus Bomb: The Manhattan Project and Government in the Dark* (2016). The threat of nuclear annihilation is a result of science. Science is a gunslinger for hire in many ways. Scientists may be motivated by great and admirable goals of assistance to their fellow humans but that does not mean that they do not unleash awful destructive forces. This is not the fault necessarily of individual scientists but the result of institutions within which they operate. The Patent system for example grants a monopoly for 20 years for the inventor who reveals through publication their method of invention if new and original and representing an advance in the state of the art or known technology at the time. This creates a massive incentive to explore every avenue and to follow lucrative paths first. Race to disclosure privileges invention over precaution. The military-industrial complex is one of the other greatest forces for technological development. It is therefore no surprise that when scientists are employed in such domains the fruits may be poisonous. Neither is there anything in the inherent nature of science that imposes an ethical brake. In fact science has been weaponised to deconstruct the apparatus of meaning and morality in many senses. The dominant, practical force or a Faustian, Promethean, Luciferian, Satanic or Ahrimanic quest for knowledge, triumphs over any other consideration. The incentive to promote such an attitude is part of the scientific endeavour itself.

The modern manifestations of scientism derive from a long and respectable intellectual history. Unfortunately, the cumulative effect of technology and the combination of particular insights brought together by an attitude of scientism, produce an effect greater than the sum of its parts. A sense of superiority and certainty seeps into a disposition that seeks to displace all other pretensions to understanding the world betraying a closed mind inconsistent with the ostensible spirit of science itself. The scientific trajectory from Bacon through Galileo, Newton, Diderot, Descartes as well as Linnaeus, Comte and Henri de Saint-Simon creates a strong lineage. De Mettrie saw man as a machine. The impact of Darwin and the advocacy of Thomas H. Huxley continue down to Julian Huxley who was a modern articulator of the term transhumanism. Claims to rationality, reason, empiricism and positivism have assumed an authority similar to the certainty claimed by conventions science criticised and despised. Popper, Quine and others identify shortcomings in the method. People such as C.S. Lewis, Manly Hall, Patrick Wood and others have seen how this stream has become poisoned by its own potency.

The scientocracy which C.S. Lewis feared was proposed in the form of technocracy after World War I by people such as M.K. Hubbert in the *Technocracy Study Course* (1934). The conception was of a team of experts such as engineers who identify requirements to manage a command economy based on the model of co-operation utilised during the First World War. Such technocracy requires a close command of the energy system. Close surveillance of the citizen would occur in a context where private property did not exist. The apparatus of this brave new world order would require a lot of data but they knew that such management systems would evolve eventually. Independence and democracy would disappear.

In the 1920's certain scientists pointed to the future. Essays like 'Daedulus, or, Science and the Future' (1923) by J.B.S. Haldane foreshadowed the scientist as a superman beyond the superstition of morals applying their knowledge,

> *"...conscious of his ghastly mission and proud of it."*

Bernal just a few years later laid it out with prescience and no pretension of hiding the ghastly mission. Technocracy and technetronics are more manifestations of the totalitarian-tending thoughts of J.D. Bernal and his ilk. He indicated the scientocratic, autocratic dream, making clear to me that it would be a nightmare for us outsiders. Furthermore, in his work on the 'science of science,' he indicated an effective model for regimes such as contemporary China. In his essay *The Social Function of Science* (1938) which became a book, he argued,

> *"Already we have in the practice of science the prototype for all human action. The task which the scientists have undertaken — the understanding and control of nature and of man himself — is merely the conscious expression of the task of human society."*

Bernal argued that science is communism. It is no surprise that his 'scientometric' and 'science of science' ideas has had a huge influence on the scientific-technocratic fabric of China. He also wrote *Science in History* (1954).

It is clear from crucial figures such as Bernal, that science, technology and scientometrics are a deeper enterprise than objective study. Science represents a totalising or rather a totalitarian ideology that provides answers and through a science of science allows control effectively of all significant knowledge. Implicit in that is a dominant quantitative endeavour. This comprehensive view of science is capable and destined to control nature, society and humankind itself. The expert, scientist or technocrat combine through their shared appreciation of this exclusivity and comprehensiveness of its power to control society. Capitalism and communism are merely a materialist pincer movement to this final reality.

The impetus from Huxley through Haldane, Wells, Bernal and others, facilitated by concepts from Bernays in relation to propaganda and complemented by cybernetics, fit in with ideas of technocracy that grew up in the US and also manifested in notions of a 'technetronic' society. The technocracy or technetronic society as promoted by Brzezinski in *Between Two Ages: America's Role in the Technetronic Era* (1970) works equally well in China and the West. Manifestations of concern about emergent scientocracy are evident in the rise of the *Society of the Spectacle* (1967) by Debord and surveillance as explained in *The Age of Surveillance Capitalism* by Zuboff (2019). Massive economic rewards and institutional capture by commercial interests with our failure of critical thinking and promotion of philosophies consistent with scorched earth scientism have emboldened participants and proponents to seal the final deal and project their vision onto us for our own good. A technocratic dream which is a nightmare for the mass of humanity will be sold on fear of environmental dangers by people whose contributions are inflated when successful and ignored when unsuccessful.

This is no conspiracy theory, but accurate description of the existence and nature of pervasive scientism as a totalising and authoritarian force without any opposition using the battering ram of technology as proof of power. By definition, opposition must be framed in scientific terms and presumably even so would only lead to an ultimate incorporation into the machine of science or rejection as being non-scientific. We are to leave management of consciousness and true self to people who do not believe in one or either. We are to imagine that we will be reproducible in some non-carbon form by people who do not see anything beyond a reactive machine in the body. We are to yield control of our cortex just because some want to develop technology. Like Hansel and Gretel we will wonder at the candy house and be taken in by promises. We will be told that people who deny much of what we value intend to create a utopia for us, a cornucopia of having whatever we want from immortality to supermanhood. Trotsky even sells this idea. Whether globalist fascism or communism, it leads to scientocracy. A spurious Green Scientocracy is a real possibility.

Tyranny will not be established by scientists driven by the best of science but by opportunists extending its worst. Arendt examined totalitarianism and the use of terror by 20th century regimes of the left and right. She wrote *Eichmann in Jerusalem: A Report on the Banality of Evil* (1963) on the machine nature of such systems. Lasch wrote *The Culture of Narcissism* (1979) and *The Revolt of the Elites and the Betrayal of Democracy* (1994). He anticipated a bureaucracy filled with narcissists with a sense of omniscience, superiority, shallow affect, a sense of right to exploit others and sadistic elements. These reflect a wider culture of youthfulness, afraid of death, unconcerned with posterity, distorting time, consuming now, seeking material success, focusing on appearance to others, stimulating infantilist oral cravings and lacking inner direction. It seems that certain pathologies such as narcissism, with its manipulative nature, are ideal for contemporary bureaucratic success of those who crave feedback for their sense of self. The machine will be fueled by an imperial governing class imbued with narcissism, managerialism, expertise, gadget wizardry, technologism and the usual acolytes.

Critics have long perceived that the pendulum of policy pushed by science has swung to another extreme in opposition to religion and spirituality. Positivism, empiricism and materialism made magical means of machines. The machine of machines with management of science has become a monster which will consume not just the individual but the race itself as we know it. Excessive belief in potency of such ostensibly sensible doctrines persists despite profoundly negative effects thereof manifest in mechanisms of conflict, war, environmental degradation, unethical experiments and human alienation. Having yielded sovereignty, seduced by hypnosis of mass mainstream media based on propaganda and being mesmerised by tunes piped at the behest of puppet-masters of power, we face tyranny. That tyrannical tendency has an appearance of reason and appeals to supposedly inevitable social factors, indisputable measurements and quantification with the soft voice of sense unimpeded by ethics save evidence about realisation of some pressing objective. Hayek warned in *The Counter-Revolution of Science* (1952) how Hegel, Comte, Feuerbach and Marx produced totalitarianism.

Tyranny is a ubiquitous tendency in human affairs. The difference is that now technology complemented by pharmaceuticals allows servitude to a degree never available hitherto. Many thinkers have been aware of the threats posed by technology. In the triangulation of conceptual terminology emergence between psychedelics, transhumanism and AI in 1956-1957, conditions for future slavery could be seen. Globalisation was a tactic rather than a strategy. The strategy was control and termination of homo sapiens as we know it. Writers like Philip K. Dick began to perceive what was happening. The novels indicated a future of fake realities, alienation and dull dictatorship. The Empire of Scientism is based on a coalition of ideas, forces and tools that combine players in various games into an orchestration of all people. Conspiracy simply refers to an agreement, express or implied to achieve some goal. We can trace it to people like Adam Smith in *The Wealth of Nations* (1776). He said that people of the same trade seldom meet even for merriment or diversion but there is a 'conspiracy against the public.' With hi-tech convergence is convergence of industry which invites and may reflect conspiracy. Conspiracy in legal terms is often an agreement by nod or wink between different parties and does not require express or even secret engagement in certain cases. There is a conspiracy to rule with technology and an associated one to change the human race irrevocably.

Away From Dispiriting Conspiracy

"...I have given some idea of one way in which such development could take place by the colonization of the universe and the mechanization of the human body."

Bernal, J.D.

The tendency to describe any unwelcome view as a 'conspiracy theory' is now pervasive. The term 'conspiracy theory' is often used as a method to discredit arguments, propositions, discussions and concerns and ignore the possible merit that may underpin or motivate the original concern. At the same time, the ones who employ the concept of conspiracy in a most effective way are Government authorities. The English Government has a long history of uncovering conspiracies at very opportune times. Whether they were always genuine or not is another matter. If they were not genuine and rather staged in some way, then it ironically emphasises the reality about knowledge of conspiracies as largely false by those who know best. But independently of speculation it is clear that the State uses conspiracy as a very real notion.

In the US, there is enshrinement of conspiracy in the Sherman Act 1890. The State acknowledges conspiracies exist and should be punished. Why are such conspiracies criminalised? The idea is that agreement may manipulate the market to the detriment of the consumer. The Supreme Court has wrestled with this issue. The domain of competition law or anti-trust is intended to prevent market abuses. In addition, many notorious criminals will be prosecuted with conspiracy and often in circumstances where evidence is weak or circumstantial. So to say that conspiracies do not exist is foolish. Much security and intelligence work involves conspiracy however justifiable or not. This was part of the idea of Chesterton's novel *The Man Who Was Thursday* (1908) where nearly all the anarchists were undercover State agents. If a conspiracy is secret, it is no surprise that we must propose a theory of its existence based on circumstantial evidence sometimes.

Yet there is an irrefutable conspiracy which is described as an 'open conspiracy' explained by the greatest proponent of scientism - H.G. Wells. It refers to a conception of world government by scientists. Wells wrote a book explaining existence of this conspiracy. He also wrote a book explaining that this conspiracy was 'The New World Order.' It corresponds with Bernal's insane plan. The Wellsian written confession is clear and proves beyond a reasonable doubt the reality of the intent. Nevertheless, to say that the world will be run by scientists seems to invite accusations of paranoia or at least a 'paranoid style.' This accusation applies to scientists also. John C. Lilly not only made contributions to spiritual evolution but he also worked on technologies and studies that were about brainwashing and control of human behaviour in the 50's onwards. He saw possibilities and dangers. He then believed that machines we were creating would control us and he was in quite a good position to so propose. His work involved insertions into animal brains and thus represented exploration of a technology that corresponds with transhumanism. Science has been studying behaviour in order to control it.

> *"Psychological and physiological discoveries will give the ruling powers the means of directing the masses in harmless occupations and of maintaining a perfect docility under the appearance of perfect freedom."*
>
> Bernal, J.D.

One obstacle, in the path of those advocating advance from ostensible trapping in flesh towards 'transcendence' through technological enhancement is the spirit superstition. That is how scientism sees it and why erasure and spirit denial is part of the Empire formation process. Spirit is treacherous to the advent of scientocratic utopia or E-topia. The attempt to dissolve the idea of spirit has been ongoing from the Enlightenment, although the Reformation may have started the process. Excising spirit from discourse became a constant consideration in central ideas of rationality and reason in an emerging, enlightened, empirical Empire. Notice the similarity between 'empirical' and 'empire.' The empirical element is a reason for its success when used correctly in science but in error when extended inappropriately. The 'discovery' of the unconscious and subconscious were consequences of deconstructing the concept of spirit. It was not that we did not have these things but they were called something else. Thus it was like the fiction of America's 'discovery' by Columbus. Things got more troublesome for spirit-denying science by its necessary substitute phenomenon of consciousness. Spirit fit the bill and described fairly accurately on the tin what it was. Spirit was that mysterious, invisible force that enchanted the universe and manifested like wind or breath. It was the description of our essence usually conceived as being substance of the universe or Great Spirit or Creator however one might conceive that. Sometimes the concept of soul was used as co-terminous or synonymous. Spirit represented the totality of our conscious experience and was subject matter of much debate. While spirit differs in description say from Christianity to the East, and while notionally rejected in some traditions, the cohesive and persistent entity or something analogous was at the base of survival after death ideas and a domain subsequently occupied by parapsychology. The latter sought to use science as a bridge to facilitate dialogue between two conceptions of the world. While scientists often claim that spirituality threatened the existence and practice of science, the truth is that the latter practice has more often sought to uproot the former. The tradition of philosophical idealism may also address this spiritual force more in terms of a mental one.

Science escapes scrutiny or the necessity to explain or justify its direction whilst demanding it from others. We will not hear much mention of Unit 731, nor Hiroshima, Nagasaki, Fukushima, Chernobyl, Dr. Mengele, Thalidomide nor the Tuskegee Syphilis Experiment. We will not hear much about professors who helped with human zoos in the nineteenth century. The Vipeholm experiments will be glossed over. This may be because there is no single institution representing science and because it is part of the continuing and future apparatus of control. Criticising dangers from establishment of institutional hegemony and oppressive systems we must be consistent. It is easy to jump from the frying pan into the fire. That an institution was too powerful and did terrible things is no argument in favour of sacrificing power to another institution or practice with similar ideological certitude. Lessons of such events and histories can be utilised for other power consolidation. Thus the Catholic Church is being accused of 'spiritual terrorism' by some. The danger is that all negative activity associated with institutions which manage some spiritual activity are imputed to all people who merely engage in any spiritual activity and have even rejected institutional religion. That is part of the danger of constant creation of a materialist narrative which may despise anything associated with spirit and seeks to impute guilt by association with some nefarious practices. Intent to uproot spirit as a force that impedes full efflorescence of a materialist, scientific, technocratic determinism and totalitarian ideology should not be underestimated. That one autocratic system has engaged in wrongdoing does not provide adequate basis for justification of an alternative, potentially even worse authoritarian substitute.

The conspiracy is so clear that it was labelled 'an open conspiracy' by H.G.Wells. His blueprint for a World Directorate with collective biological controls triumphing over individual rights and religion in a scientific society is manifesting. Spirituality must perish as an impediment to final resolution in favour of scientism. Wells was more of a literary man than a scientist. It is only by awareness about the agenda and associated counter-action in favour of individuality that a counter-force of moral authority may save the spirit and the human race. Otherwise the conspiracy to create an Empire of Scientism becomes inevitable and its inherent composition will be based on compulsion and a reign of terror posing as sense.

So conspiracy must always be regarded as a threat in reality where people have any incentive to conspire. Such conspiracy as exists in the sense acknowledged by Wells as well as any co-ordination that exists through co-operation to achieve technological networks create dangers of abuse. The absence of international regulation creates a space wherein informal networks may make agreements to achieve objectives particularly if related to the transnational sphere beyond democratic and national control. Technological networks invite conspiracy or agreement between a number of players. Any such conspiracy must be superficially dispiriting insofar as it threatens to remove the benefits of competition for consumers. Co-operation between players in a game removes doubt and destroys competition. But there is another way that the converging, centralising, globalising tendency of tech-networks facilitated by the class of international experts, technocrats and bureaucrats, in a mood of Narcissistic movement and kept together by the ideology of scientism, may represent a dispiriting conspiracy.

Dispiriting may refer to the process of becoming disappointed, disillusioned and disenchanted. This has been the effect of Enlightenment going back to elements in ancient Greece and not contained in the particular historic period most associated with it. But I suggest that dis-spiriting also refers to a process that began in the West probably at the time of the Reformation and continued during the Industrial Revolution up to the present. *Dispiriting is an active attempt or policy to banish the notion of spirit from scientific and public discourse to reduce competition and achieve intellectual dominance.*

In many ways this is how spirit became a 'ghost in the machine' to borrow a phrase that Gilbert Ryle used in *The Concept of Mind* (1949). The spirit was marginalised and gradually excised. Scientism cut out the spirit. This can be seen most clearly in the mid-nineteenth century in the UK. Howitt's *The History of the Supernatural in All Ages and Nations* (1863) occurred in a time of challenge to Christianity. His work is a defense of the Bible but also recognises that the challenge being mounted by science meant the supernatural and spiritual must be totally denied. That this was the object of scientism can be seen in the work of Thomas H. Huxley and the X Club that he formed at that same time. His band would control the institutions of scientific learning with a radical denial and a rejection of the spiritual. As the spiritualist movement came to the fore, Huxley made clear his distaste and indicated that he would not be interested therein even if proven. In this context parapsychology began and in the start of the 1870's investigators like Serjeant Cox proposed the use of the term 'psychic forces.' The growth of terms such as psychic forces and psychology was an attempt to appease a hostile scientific movement. Psychology and psychic forces were efforts to replace the spirit having flown. Psychology was a cuckoo's egg laid in another nest. Anthropology became important here and some anthropologists attended spiritualist demonstrations but retreated to scientific hostility. That they were supposedly neutral observers who could appreciate the spiritual beliefs of others was not the full story. The truth is that these people were intoxicated with an illusion of supreme knowledge and certainty. They were part of a great Empire that spanned the globe and ruled people with a sense of superiority using superior technology. As the British Empire and others declined, that attitude was transferred into tools of science and scientism. The intellectual vanguard imbued with this right to rule the world sought to substitute militarism with management through scientism. Psychology became a useful part of the apparatus of behavioural study which plays a crucial role in cybernetics. Cyberspace and cybernetics unite in ideas of 'cybernetic space' which could be used to describe the technosphere to be inhabited by conscious agents of which we are but one.

Elevating individual souls or spiritual evolution was replaced with a paradigm based on predictability and management. Out of this environment, that Huxley and others fostered, rose Marx. To the area of London round the British Museum later came Lenin and Trotsky. Trotsky would celebrate the future superman informed by technology. In the same area H.G. Wells would propagate his imperial scientism conspiracy. There too Bernal would later argue that science was communism and scientists should take over the world. In that decade Gramsci in Italy wrote *The Prison Notebooks* and realised communism needed to take out Catholicism or their empire could never succeed. The Empire of Scientism whether left or right, needed to take away religion and spirit which lay at the base of those beliefs in order to come to power. The destruction of religion and spirit would remove any impediments to full control in a totalitarian system run by those who shared the ideology of scientism. Then the true masters could assume their magical kingdom and indulge themselves in any experiment that tickled their fancy. We will be ghosts haunting the machine.

Towards the Animasphere

This is not a charter of despair, nor a call to arms but a rally of individuals. There are four elements to the start of the solution.

a) Spiritual or Personal Evolution.

At base it is a question of spiritual evolution. We are beings with spirit. Our spiritual consciousness is what defines us. For some those terms are difficult, so if one cannot accept 'spirit' they can substitute a word which represents their highest executive and noble essence interested in love, freedom, beauty, goodness, truth and the highest emotions and states. The spirit must grow through self-realisation and transcendence with respect for other spiritual beings and life. We also operate in an integrated, interdependent ecology whose flourishing is critical for our well-being and existence. We will not survive without love and appreciation of our physical world and our bodies. We will not survive without allowing habitats and life-forms to co-exist and prosper. We will not survive without protecting and preserving fresh water. If we look adequately in the proper places we will see that perennial philosophy and theology is not hostile to the earth and sees a divine immanence that is often omitted when power structures obscure it.

All great mystics talk about spiritual evolution. It is our individual necessity and that is the base of community evolution. The shaman works for the community. The hero brings the boon back to it. The hermit should bring the light back from the cave. As Evelyn Underhill points out in *Mysticism* (1911), mystics return with something. As Maslow points out in *Religions, Values and Peak Experiences* (1964), the self-actualised person who has attained peak or plateau experience will make the highest contribution to society. Not everyone has to be Mahatma Gandhi and the householder is just as significant as the monk. When the individual lives with their own true self and employs their own consciousness, they access the greatest force in the universe.

b) Resisting Colonisation of Mind and Spirit.

The scientocracy that will be the Empire of Scientism will literally colonise the body, mind and thus thwart the spirit. You are the new colony, your booty is up for grabs, your treasure can be mined. You will not be merely servant or sufferant of the colonial power, you will be the colony itself to be managed, subjugated and controlled. The biosphere and ethnosphere devastated by industrialisation will be transformed further into a technosphere which takes over nature and human nature. This is what is being openly espoused in works of widespread popular appeal to a public that is either hypnotised or asleep.

We are asked to believe that the institution of science, implicated in the destruction of nature, is now our only saviour with no evidence of any philosophical shift in its deadly worldview other than standard self-interest and opportunity. The production of inherently dangerous technologies continues with an ostensible recklessness to obvious risk and a willingness to throw caution to the wind in satisfaction of short-term incentives. In the long term perhaps we must presume scientists calculate consciously or unconsciously, that their skills will be needed to clean up the mess they have created and can worry about consequences afterwards and the devil may care.

> *"The world might in fact be transformed into a human zoo, a zoo so intelligently managed that its inhabitants are not aware that they are there merely for the purposes of observation and experiment."*
>
> *Bernal, J.D.*

It will go like this. You will be told daily that all the past was terrible and only present progress is the way forward. Spiritual terms will be appropriated and inverted. Slogans will be pounded into your nervous system using the electronic system you are already hooked to. Science is your saviour. Technological progress is inevitable and inevitably beneficial. Trust the experts. All who oppose this are ignorant savages who want to 'run around naked.' The language of colonisation is close to this enterprise because that is what it is. This is the empire that uses scientism as imperial force to exert dominion over the mind and spirit of humanity. It is destined to be reduced by the imperial cybernetic, AI, algorithmic network, turning people into cyborgs as a part of the transition. This transhumanist phase is trance-humanism. The public are hypnotised to believe in the inevitable end of humanity through technology. The public will be promised great wonders and powers in return for giving up democracy as proponents from Trotsky to Wells have always done.

Encouragement and seduction will yield to compulsion once networks are pervasive and irresistible. Some type of humanity or de-transitioning will be permitted by those who advocate that we be subsumed in ineluctable subjugation or submission to 'superintelligence.' In the transitional phase you will be told of the great possibilities of enhancement. You will be subtly and then more severely compelled to co-operate in becoming cyborgs. Opponents will be hounded and attacked for crimes allegedly committed by them and their supposed ancestors. Groups will be attacked regularly so that possibilities of resistance are avoided through the poison of suspicion. Civil war could be fomented to create the self-fulfilling prophecy that all spiritual belief is responsible for all the ills of humanity hitherto. In the meantime, the intellectual substitutions and sleight of hand will continue. Spiritual terms will be appropriated. Simulacra of spiritual experience shorn of any normative or dogmatic content will be presented as genuine so that the substance is siphoned from the body politic. Bodies presumed to be concerned with spiritual issues will sacrifice their supposed congregations to the new control system. A carousel of inflated crises or draconian responses to actual ones will cause confusion necessary to create an appetite and acceptance of greater control. A de-moralised and de-humanised population will dwindle so that all that is left is a shell necessary to ensure that the system persists and to allow experimentation for the elite. Thus, intellectual awareness, threat identification and resistance are a necessary strategy to combat scientocratic advance.

c) Being Pragmatic and Cosmopolitan in favour of our Spiritual Nature and the Spirit of Nature.

We must be pragmatic in approach and cosmopolitan in attitude to respond to challenges and technocratic dangers. We could look to the work of William James here. We should resist manipulation that operates to reduce spiritual diversity and to disrespect the very existence of traditions. At the same time people and practitioners must be able to transcend narrow confines of an apparatus of belief systems that may only serve our enslavement. Co-operation consistent with the perennial philosophy and even theology that recognises the primacy of spiritual life should occur above petty restricting belief. No religious system can afford to insist on the comprehensiveness and exclusive purity and truth of its own message and methodology in order to avoid co-operation without risking its very existence by that failure to engage and co-operate and resultant defeat of the ultimate aspirations of its endeavour. Religious leaders have largely missed the seriousness of an existential challenge to the enterprise they claim responsibility for.

We must avoid undue condemnation and criticism of people who may act inconsistently with some of our beliefs insofar as objectives of co-operation and common interest in protecting the very nature of the ethnosphere, biosphere and the spirit of humanity itself are threatened by the monstrous technosphere and its digital network developing to condition and control us and finally eradicate spiritual possibility. The universal sphere of spiritual comprehension of humankind including all the created or evolved domains or dimensions might be called the animasphere. Anima here refers to the spirit, soul, life or inner self. There will be no sustainable biosphere, ethnosphere without this animasphere to balance the terrible threat of the technosphere.

d) Conceive the 'Animasphere.'

Strategies of non-violent, spiritual, moral and mystical force utilising innovative and traditional approaches may address the threat. Invoking a metaphorical murmuration of mystical and spiritual force consistent with perennial philosophy past and present and pluralistically drawing from core traditions can help counter-balance this transition. Mass spiritual solidarity is necessary as servants of the new scientocracy may promote splintering in potential sources of opposition. Perhaps inspired by the invisible congregation of light that Eckartshausen talked of in *The Cloud Upon the Sanctuary* (1791) we can act in our own individual domains, beyond possibility of corralling and capture. Persuasion of players in the technological world can assist. The network of pervasive surveillance and supervision makes rebellion difficult. Violence is not the answer. It is the battlefield of attitudes, psychology, perceptions, strategies, tactics, philosophies and ideas whereupon we must engage unarmed save with conviction and certainty of the true nature of human consciousness or spirit.

The 'imaginal' as used by Henry Corbin was also used by Frederic Meyers before and Bruce Lipton after. The imaginal in scientific terms indicates that cells in a chrysalis resonate individually, and operating within the same DNA, transform and evolve out of that form to another. This is a model for mystical possibilities that will defy capture, containment and constraint as all oppositional movements will experience. It is not violence that will be the way forward but imagination and love sharpened into myriad points of counter-active moral force.

Tolkien believed that The Ring represented power, magic control and the machine. He seemed to believe that if Gandalf got control he would be worse than the bad guy, believing that he was using power for good ends. C.S. Lewis had a similar belief that the worst tyranny would be one supposedly exercised for the good of its victims by convinced do-gooders acting in accordance with their conscience. Science has mutated into an Empire. The Empire of Scientism is the new empire of the mind foreseen by some. The apparatus of empire has already been turned to new colonies. It is not just existing or future colonies on other planets but the last colonisation will be that of human mind and spirit.

Pamphlets may be against revolution such as that of Burke in *Reflections on the Revolution in France* (1790). Pamphlets may foment revolution such as Paine's *Common Sense* (1776). Both emerge from turbulent times where the inevitable tensions must be reconciled. I believe that we can peacefully achieve conditions that avoid the worst of repression while addressing the aims of reform. However, dunes in the scientism desert creep over all. The domination of existential thought, material and mechanistic mind-sets and exaggerated claims of empiricism, positivism and scientism move relentlessly. Marram grass is a plant that is familiar to many who walk by the sea. The network of stalks is able to contain the spread of sand. Individuals can only form a persistent network when they establish common cause. Such cause must come from individual realisation informed by one's own experience and observations. That is true for radical empiricism as promoted by William James. Mass movements fail because they are usually captured and channelled. Rather individuals must be magnetised by becoming disposed of their own volition. A mass movement magnetised by proper perception or analysis can respond and correspond in a unified way that cannot be thwarted. Elites can be opposed by masses peacefully and pragmatically through mass awareness. Distribution of people in a mass movement not confined, contained or constrained through an organisation capable of capture, creates conditions necessary to provide feedback, revelation, information and a true picture of reality. The occult becomes revealed when millions of pinpricks of light produced by many individuals are motivated to puncture the matrix of deception.

This is a commitment to optimism and light in a dark time, to persuade rather that persecute. I reject notions that any one group or class is responsible but a concatenation of forces produced by the cosmopolitan interplay of various functions. Science is not the problem. However, the goal of scientism calculated to produce a scientocracy to create a material utopia through open and clandestine conspiracy, as revealed by Wells, provides the real coherence of the coalition. In my novel *Blue Lies September* (2019), I posited a draconian State in the UK based on opportunity provided by a genuine crisis before the present one. I speculated on dystopia in *Ireland 1*

Don't Recognise Who She Is (2019). In January 2020 in an episode about 'Globalization' on New Thinking Allowed (published in March on YouTube) I indicated my fears of global control under the guise of scientism. I think my views should at least be seriously considered. It is now hard to conceive dystopias when reality outpaces depressing speculation daily. There are no demonic or alien forces utilised here. Instead it is the same litany and human nature tragedy anticipated in legends of Narcissus, Icarus and many more. But fate is not determined.

In summary, science may be regarded as an ideology and worse still scientism definitely is an ideology with more inherent dangers. Scientism creates the illusion of certainty or truth that acts as a coherent force to unite materialists, scientists, entrepreneurs, capitalists and communists in a dogma or ideology tending to ideocracy or scientocracy. Scientocracy is an advanced form of technical control by technocrats based on establishing objectives deriving from scientism, scientific self-interest and an associated class of entrepreneurs and managers. Scientocracy will be totalitarian in the highest degree. Scientism reflects an almost religious status of science, machine-worship, tolerance of dictatorship, reliance on instrumental technical solutions and experts, over-reliance on metrics and quantification and a refusal to adopt holistic analyses or solutions. Science, scientism and technocrats promote a reductionist view of humanity and seek to deny very important elements such as rights, personhood and human dignity itself. Stressing the exponential nature of technological growth promotes transhumanism as inevitable and suggests that homo sapiens must yield to mastery by machines. This reflects the idea that we are machines, that society is a machine and that we will become machines or slaves.

We have seen how government has worked within capitalist and communist systems to create societies of spectacle and surveillance. Surrogates and sycophants to such forces promote erosion of rights while expanding coercion and compulsion. Conduct of our lives is being constrained by systems supposedly constructed for convenience. We are governed by the unelected. Our deepest values are being replaced by shoddy substitutes and simulacra. Our spiritual lives are being siphoned and meaning is sucked out. We are threatened by human-species-tech-suicide through transhumanism. We are imprisoned by mathematical models. We are increasingly denied access to nature despoiled by our supposed saviours who prepare to abandon us. We are being conditioned to live in a technosphere before our demise. C.S. Lewis in his novels and non-fiction work anticipated scientocracy, misuse of public resources, manipulation of the academy, concocted news and a deadly conspiracy of scientists being magician-materialists. Unless the mystical, indigenous, spiritual, ecological and properly scientific unite, balance will be lost and disaster inevitable. To elevate scientific method to absolute global governance is the ultimate insanity resulting from scientism. Conquer your fears once and for all or continue to be conquered forever.

If we do not recover our autonomy and authority over our own life then an apparatus of authoritarianism will grow, even if administered with pen and paper. We seek external affirmation instead of inner direction. We have lost confidence. We have chosen anxiety and substituted experts for community. We have alienated ourselves from ourselves and our spirit. We have ceded sovereignty and conceded our ability, desire and right to control our own lives. We have been convinced that objectivity manifested by authority with attitude and figures can figure out better than us who we are and what we should be. Unless we rediscover and cultivate our sovereignty and most of all our spiritual consciousness, we will lose to a totalitarian Empire of Scientism that moves from appearance of concern to apparatus of heartlessness and terror convinced by a righteousness in its ability, unconcerned by its subject's needs.

It may also be that we experience some resistance to centralisation that manifests in a universal civil war in the Western world and beyond and perhaps widespread opposition in the Islamic world. Another possibility is that a zealous pursuit of scientific rigour creates a Scientific Reformation to rid the world of superstition and many practices that might be deemed non-scientific. That is why caution should counsel those who seek to expedite an Enlightenment reign in an Empire of Scientism. The World Bank is forbidden from concerning itself with democracy. That technocrats might ignore basic human rights is no surprise in contexts where liberty and rights are not express concerns for them. As Easterly identified, free development is far superior in pragmatic terms than authoritarian development.

People who believe in our spiritual or transcendent dimension should enhance their co-operation as a matter of urgency to combat the tech-tsunami and scientism surge that will drown all spiritual enterprises with its force. What I call the 'animasphere' could be employed here.

Animasphere indicates a domain of perennial wisdom, cosmology, theology, metaphysics, philosophy or belief in divine, supernatural, supra-sensible and spiritual reality and an inhabited, animated otherworld being at times humanistic, poetic, mythic, legendary, animistic, unconscious, subconscious, imaginal or transcendent and timeless, often studied by folklore, parapsychology and theology hitherto and although mystically accessible seldom reducible by limited scientific tools.

The animasphere would include our universal spiritual heritage of the human race conceived in a cosmopolitan way that complements ideas about the biosphere and ethnosphere. The concept suggests the basis for a re-alignment with earth and environment, informed by inputs of indigenous spirituality and doctrines discarded by some theological traditions. It promotes the idea of respect and does not insist on exclusive truth to the detriment of other spiritual paths. A thriving zone would require persuasion rather than force for faith proponents. Promotion of a healthy animasphere could be an antidote to the technosphere's tentacles facilitated by perceived self-interest of those who practice versions consistent with the perennial philosophy or theology. If exclusivity and judgment is laid aside, then internal, internecine division may be avoided and combination may renew strength in spiritual health of the human race.

Otherwise it will be globetechgov, scientocracy or scientocrazy. We face a scientocracy based on scientism that will sever our connections with spirit and nature. The illusion of efficacy in relation to technocratic intervention, central planning and assumption of some autonomous reasonable force to enforce it, creates what Hayek calls 'collectivist hubris' of conscious direction. The Empire of Scientism will be an inevitable consequence of ceding our spiritual sovereignty and individual critical faculty to authority. This Machine is fueled by you, no longer a mere citizen or subject but a thing or object subjugated to its monstrous power without rights, personhood, identity, free will or even humanity. The world you live in is ruled by a human otherworld as it is. Meetings in basements, informally and formally, at strange locations, in secret or openly, for specific or general reasons, by administrators, entrepreneurs, political players, opinion makers and most of all scientists, constitute the source of actual and potential control of your future and that of your species. Most of these meetings are dominated by a worldview that rejects spirituality and are largely beyond public scrutiny.

The insight of Günther Anders and others was that we are now living within a technological world and our lives increasingly happen within it. We are becoming aliens in a technosphere where other conscious agents will rule us. Anders indicated a kind of alienation and a strange shame in comparison with machines. This is because many have convinced themselves that we just happened by accident without purpose and are in some way inferior to that which is planned and has objectives. The slightly better explanation is that we have invested our resources in people who desire to achieve magical power at our expense. This is often based on pathological rejection of any spiritual quality and increasingly physical and human ones as well. Day by day we ourselves bow down and adore gimmicks, gimcracks and geegaws. We spend time on puppet-shows and cave shadows and give up our spiritual, political and biological sovereignty.

Scientific propaganda has been so successful that the word 'spiritual' is difficult for many who are sympathetic to the critique of scientism. There is no necessity to require anyone accept ideas like the animasphere or concept or reality of spirituality in order to share the critique of scientism. There is some awareness in the humanities that a colonial expeditionary force sailing on the 'consilience' ship may be an effort to take over the humanities by the scientism endeavor. Colonialists will expand. At some stage the humanities will realise that concepts of humanity are on life-support. They will understand that they cannot proceed when perennial wisdom, that was the foundation of humanism, is undermined. However, realisation may come too late. Meanwhile therefore the issue could be looked at as one of methodology going back to philosophical work like Wilhelm Dilthey. The issue then is about distinctions between the inner and outer. Again, scientism is blurring this dichotomy and making methodological solutions less workable. Scientists that focused on evidence and reason properly like Emil du Bois Reymond in a famous speech in Berlin in 1880 argued that certain fundamental issues could never be answered by science. We will never know. Rudolf Steiner agreed with this awareness about limitations of scientific knowledge and proposed a separate domain of spiritual science. Nevertheless he saw materialism and modern science as an Ahrimanic illusion.

If you experience a profound shock soon, perhaps when you perceive the plethora of devices, machines and agents in the hi-tech agenda, you may see it. If you think that scientists and technocrats have nothing new in store, you will be surprised. If you do not realise the implications of extensive robots and pervasive 'love dolls' perhaps you will shortly.

This work was called a pamphlet. There are international definitions of what a pamphlet is although they are not definitive. They are often defined by length and whether they are bound or not. Thus some will object to my description. The loose leaf is impractical. Length depends on size of pages. The real significance of pamphlets in political or sociological terms was to address issues of the highest significance which weighed on the author and often transcended any particular party political perspective. Even so there are pamphlets of this length. Nevertheless, the issue of imminent imperial scientocracy is one that should be approached as a unique problem such as we have never encountered before. This work has not dwelt on bad science nor how scientists can mislead the public nor how scientists may have incentives to produce certain results and so on.

Scientism poses problems for science itself. Science is already an elephant in the room. Scientism is a mechanical or bio-engineered elephant in the room that is out of control. Increasingly you will hear arguments that appear to come from mainstream positivism in science building on theorists such as Rudolf Carnap, Otto Neurath or Auguste Comte. Such theories thought that science could explain everything and were hostile to metaphysics. The result we have to fear now is that this belief seeks to cut our more healthy tissue. It will be used to shift the burden of proof on to anyone who is unable to prove a proposition in scientific terms. Thus any proposition that has an appearance of a scientific basis however flimsy will triumph over any other no matter how deep or meaningful.

Mainstream science has issues about how it produces research results and to what extent they are compromised by funding mechanisms and can be reproduced. If the scientism ideology is enshrined in the apparatus of governance then the results desired will be created to ensure persistence of the governors whose mandate is without democratic legitimacy anyway. One of the inspirational works for scientism was *The Martyrdom of Man* (1872) by William Winwood Reade. Encouraging the rise of the rational and scientific must come at the expense of religion and spirituality, he realised. Thus he wrote that to build - we must first destroy. Bear in mind when you hear the technocrats and scientocrats talking about building or 'building back,' that while it might be described as 'better' by them - ask what it is they want to destroy or replace?

We need an instinctual prescience informed by conscience to invigilate science so the ideology of scientism does not enslave us. Causation operates in a downward way, as well as upwards. Only by navigating according to the heavens of perennial wisdom can we keep on course and ensure that the enterprise of spiritual evolution and humanity persists. Biological transcendence is a mean substitute and shoddy simulacrum for beings we are meant to be and experiences we are made for. Your mind, body and spirit will be colonised and used for the emergent Empire of Scientism. Prophets like Philip K. Dick perceived what was afoot. It has taken the rest a bit longer. We are beings of light, spiritual beings, beings with human dignity becoming ghosts in the technosphere.

And What If Then?

And what if then you found your own way?
And didn't depend on others for what you say
Or succumbed instead to your higher light play
Because as well as nod you might say nay.

And what if then your voice was revealed?
And your being within was unconcealed
Or you allowed your head or self be healed
Because to court the real heart you appealed.

And what if then you escaped the old bond?
And dwelt on that of which you should be fond
Or behold your true self in reflected pond
Because of risen eye's art to the beyond.

> Because petrified with fear we stay
> Or doubt and worry most of the day
> And what if you change the rhyme?
> And break out of the old paradigm.

About the Author

James Tunney obtained an honours degree in law from Trinity College Dublin, qualified as a Barrister at the Honorable Society of the King's Inn, Dublin and obtained an LLM from Queen Mary College, University of London.

Since then he worked as a Lecturer and Senior Lecturer in UK universities. He has been a Visiting Professor in Germany and France, lecturer around the world and worked as an international legal consultant in places such as Lesotho and Moldova for bodies such as the UNDP. He talked in many countries and published regularly on issues associated with globalisation. He has taught, written and talked about subjects such as indigenous rights, travel and tourism law, culture and heritage, IP, communications technology law, competition law, China and World Trade.

He decided to leave the academic world behind to concentrate on artistic and spiritual development. He has exhibited paintings in a number of countries and has continued his writing.

You can help my work by leaving a review on Amazon.

Printed in Great Britain
by Amazon